Quantum Mechanics of Brain Waves: A Cognitive Exploration

N.B. Singh

DEDICATION

To Nature,

I dedicate this book to you, the source of all life. You are my inspiration, my teacher, and my friend.

Thank you for teaching me about the beauty of the world around me. Thank you for showing me the power of the natural world. Thank you for giving me a sense of peace and tranquillity.

I promise to do my part to protect you and your many wonders. I will teach my children about the importance of conservation and sustainability. I will work to make the world a better place for all living things.

Thank you for everything, Nature.

With love,

N.B Singh

Contents

PREFACE

Welcome to "Quantum Mechanics of Brain Waves: A Cognitive Exploration." In writing this book, my aim is to take you on a fascinating journey that delves into the intriguing relationship between quantum mechanics and brain waves, seeking to unravel the mysteries that lie at the intersection of these two enigmatic realms. As we traverse through the pages, we'll venture into the depths of both quantum theory and neuroscience, exploring how they converge to shed light on the nature of consciousness and cognition.

The human brain is a complex and intricate web of billions of neurons interacting through electrical and chemical signals, giving rise to a wide array of brain waves that accompany our thoughts, emotions, and perceptions. Over the years, scientists have made remarkable discoveries in understanding the various brain wave patterns, correlating them with different states of consciousness and cognitive processes.

At the same time, the field of quantum mechanics has revolutionized our understanding of the microscopic world, defying classical intuitions and introducing us to a realm of superposition, entanglement, and uncertainty. Quantum phenomena have been observed in various systems, from subatomic particles to larger biological entities, prompting researchers to explore the possibility of quantum effects playing a role in biological processes, including those within the brain.

In this book, I aim to provide a comprehensive exploration of the current state of knowledge regarding quantum mechanics and brain waves, considering both theoretical concepts and experimental evidence. I will examine how quantum principles might influence neural processes, the potential implications for our understanding of cognition and consciousness, and the ongoing debates surrounding the quantum brain hypothesis.

I have written this book with both curious minds and experts in mind. For readers new to quantum mechanics or neuroscience, I offer a gentle introduction to the relevant concepts to facilitate a deeper understanding of the subject matter. To those already well-versed in either field, I hope to spark new perspectives and encourage interdisciplinary dialogue.

I would like to express my gratitude to the researchers and scientists whose groundbreaking work has paved the way for the fascinating insights presented here. Their dedication and passion for unraveling the mysteries of the mind have inspired me, and I hope this book will continue to inspire future generations of researchers.

Lastly, I acknowledge that the field of quantum cognition is rapidly evolving, and new discoveries may emerge even as this book reaches readers' hands. While I have taken great care to ensure accuracy and completeness, I encourage readers to explore the latest research and developments in this ever-evolving field.

Thank you for joining me on this enthralling journey into the realms of quantum mechanics and brain waves. Let us embark together on this cognitive exploration and, in doing so, deepen our understanding of the fundamental nature of human consciousness.

Happy exploring!

N.B Singh

Chapter 1

Introduction

1.1 The Interplay of Quantum Mechanics and Neuroscience

The remarkable union of quantum mechanics and neuroscience forms the foundation of this captivating exploration. Quantum mechanics, a revolutionary branch of physics, describes the behavior of particles at the subatomic level, while neuroscience seeks to understand the complexities of the human brain and cognition. Combining these disciplines has led to exciting hypotheses and promises to unravel the enigmas of consciousness that have perplexed humanity for centuries.

The allure of this interplay lies in the possibility that quantum phenomena might impact brain function. While classical physics effectively explains macroscopic phenomena, it might not suffice to elucidate the intricacies of the brain's microscale processes. Scientists speculate that quantum superposition and entanglement could play a role in neural systems, giving rise to emergent properties beyond classical explanation.

One intriguing notion posits that quantum coherence may exist within neural networks, enabling information processing through fundamentally different mechanisms than classical computation. Such quantum-inspired models of cognition could potentially explain the brain's unparalleled computational prowess and efficiency. Nevertheless, this hypothesis is met with both enthusiasm and skepticism, driving the ongoing pursuit of experimental evidence.

Quantum consciousness, a captivating concept, suggests that consciousness itself emerges from quantum processes in the brain. Eminent physicists like Sir Roger Penrose and Stuart Hameroff have proposed that consciousness arises from orchestrated quantum computations within the brain's microtubules. However, this hypothesis remains highly controversial, as the nature of consciousness remains one of science's most profound mysteries.

Researchers exploring the interplay of quantum mechanics and neuroscience face the challenge of bridging the microscopic and macroscopic scales. Quantum phenomena primarily manifest in small-scale systems, leaving open questions about their relevance and stability in the brain's warm, noisy, and intricate environment. Despite these challenges, recent experiments suggest the presence of quantum effects in biological processes, providing hope for groundbreaking discoveries.

Theoretical models attempting to reconcile quantum mechanics with brain function propose various mechanisms for quantum processes to influence cognition. One such mechanism is quantum tunneling, where particles can penetrate energy barriers. Researchers have hypothesized that quantum tunneling could play a role in synaptic transmission, influencing the efficiency and plasticity of neural connections.

Quantum entanglement, the phenomenon where two or more particles become correlated in such a way that their states are interconnected, also sparks interest in the context of neuroscience. The concept of entanglement has raised questions about whether it plays a role in the brain's parallel information processing and if it could explain certain cognitive phenomena, such as associative memory.

Quantum coherence and decoherence are fundamental aspects that researchers explore in the context of brain function. Coherence refers to the state of superposition, where a particle exists in multiple states simultaneously, while decoherence represents the loss of coherence due to interactions with the environment. Understanding these phenomena may shed light on how the brain maintains and loses delicate quantum states.

The Orchestrated Objective Reduction (Orch-OR) theory proposed by Penrose and Hameroff posits that quantum coherence survives for a brief period in microtubules, and orchestrated collapse events may give rise to consciousness. However, critics argue that such quantum processes in warm, wet biological systems might be highly challenging and prone to rapid decoherence.

Despite the debates and complexities, recent advancements in quantum technologies have fueled excitement in the field of neuroscience. For instance, techniques like quantum sensors and quantum-enhanced imaging offer novel tools to probe the brain's intricacies, opening avenues for potential breakthroughs.

The quest to explore the interplay of quantum mechanics and neuroscience goes beyond theoretical models and experimental evidence. Ethical considerations play a vital role, especially concerning potential applications of quantum technologies in cognitive enhancement and brain-machine interfaces. The responsible use of such advancements necessitates thoughtful discussions and ethical frameworks.

As the field of quantum cognition and neuroscience advances, collaboration between physicists, neuroscientists, and cognitive scientists becomes increasingly crucial. Interdisciplinary research provides fresh perspectives and creative solutions to unraveling the profound mysteries that lie at the interface of quantum mechanics and brain waves.

One of the primary challenges researchers face is developing robust and reliable methods to probe quantum effects in the brain. Experiments often require extraordinary precision and sensitivity, given the delicate nature of quantum states and potential noise in biological systems.

The study of brain waves, particularly through techniques like electroencephalography (EEG) and magnetoencephalography (MEG), continues to provide invaluable insights into cognitive processes. The integration of quantum-inspired approaches with traditional neuroscience methods offers a rich and exciting avenue for exploration.

In conclusion, the interplay of quantum mechanics and neuroscience is a captivating frontier that holds the promise of unlocking the profound mysteries of consciousness and cognition. Through a harmonious blend of theoretical speculation, experimental exploration, and interdisciplinary collaboration, researchers embark on an enthralling journey to deepen our understanding of the quantum nature of brain waves and the enigmatic essence of the human mind.

The famous equation by Albert Einstein relates mass (m) to energy (E) through the speed of light (c):

$$E = mc^2 \tag{1.1}$$

The Schrödinger equation describes the time-evolution of a quantum state represented by the wavefunction $\psi(x)$:

$$\psi(x) = \frac{1}{\sqrt{2\pi\hbar}} \int_{-\infty}^{\infty} \phi(k) e^{i(kx-\omega t)} dk \tag{1.2}$$

The circuit diagram in Figure 1.1 represents a simple RL circuit, where V_s is the voltage source, R is the resistance, and L is the inductance: The RL circuit

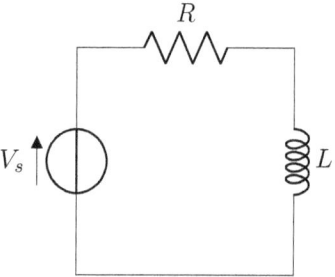

Figure 1.1: Simple RL circuit diagram.

exhibits interesting transient behavior, especially when the voltage source is suddenly applied or removed. The presence of the inductor introduces a time-varying magnetic field, leading to induced voltage and current changes. Such dynamic phenomena in electrical circuits share intriguing similarities with the wave-like behavior of particles in quantum mechanics, making the RL circuit a relevant example in the context of the interplay between quantum mechanics and neuroscience.

1.2 Understanding Brain Waves and Their Significance

Brain waves, also known as neural oscillations, are rhythmic patterns of electrical activity generated by the brain. These electrical impulses reflect the synchronized firing of groups of neurons, providing valuable insights into brain function and cognition. In this section, we delve into the nature of brain waves, their classifications, and the significance of studying them in the context of quantum mechanics and neuroscience.

The measurement and analysis of brain waves are essential tools in neuroscience. Electroencephalography (EEG), a non-invasive technique, records electrical signals on the scalp using electrodes. The resulting EEG traces display distinct patterns corresponding to different states of consciousness, such as deep sleep, meditation, and active thinking. Understanding brain waves is fundamental to comprehend various cognitive processes and mental states.

Brain waves exhibit specific frequencies, typically classified into five main categories: delta, theta, alpha, beta, and gamma. Delta waves, with frequencies below 4 Hz, are prominent during deep sleep and are associated with restorative processes and memory consolidation. Theta waves (4-8 Hz) are prevalent during meditation and daydreaming, playing a role in creativity and memory encoding.

Alpha waves (8-13 Hz) dominate when we are awake but relaxed with closed eyes. They are essential for attention and are often studied in mindfulness practices. Beta waves (13-30 Hz) are present during active thinking, problem-solving, and decision-making. Gamma waves (above 30 Hz) are associated with high-level cognitive functions and may be involved in integrating information from different brain regions.

The interplay of brain waves and quantum mechanics opens exciting possibilities for understanding cognition at a deeper level. Quantum cognition suggests that the human mind exploits quantum principles to achieve cognitive feats beyond classical capabilities. Could brain waves be influenced by quantum superposition and entanglement, enabling the brain to process information more efficiently than classical computers?

Quantum-inspired models of brain function propose that brain waves may be a product of quantum processes within neural networks. Such models could explain the brain's remarkable ability to process vast amounts of information simultaneously. The study of brain waves through the lens of quantum mechanics challenges classical assumptions about information processing in the brain.

Quantum coherence, a phenomenon central to quantum mechanics, involves particles existing in multiple states simultaneously. Researchers speculate that brain waves might exhibit coherence, allowing neurons to function coherently and exchange information effectively. Understanding the coherence of brain waves could lead to groundbreaking advances in neuroscience and artificial intelligence.

The study of brain waves and quantum mechanics raises profound questions about consciousness. How do brain waves contribute to our subjective experiences and sense of self? The Orchestrated Objective Reduction (Orch-OR) theory proposes that consciousness arises from quantum computations in microtubules. While controversial, such theories inspire novel inquiries into the nature of consciousness.

Recent research suggests that quantum effects might be at play in biological systems, with evidence of quantum coherence in photosynthesis and olfaction. Could brain waves also be a product of quantum phenomena at the cellular level? Unraveling the connection between quantum mechanics and brain waves could revolutionize our understanding of the mind-brain relationship.

Experimental investigations are essential to explore the quantum nature of brain waves. Advanced techniques, such as quantum-enhanced imaging and quantum sensors, offer exciting possibilities to probe brain activity with unprecedented precision. These cutting-edge technologies pave the way for a deeper understanding of brain function and the potential quantum aspects of cognition.

The study of brain waves has practical applications in medicine, psychology, and brain-computer interfaces. EEG-based neurofeedback, for example, allows individuals to modulate brain waves voluntarily, leading to potential therapeutic benefits for various neurological disorders.

In conclusion, brain waves are crucial indicators of brain activity, reflecting the dynamic nature of cognition. The exploration of brain waves in the context of quantum mechanics opens new avenues for understanding the brain's computational abilities and the enigmatic nature of consciousness. By combining insights from neuroscience, quantum mechanics, and interdisciplinary research, we embark on a fascinating journey to unravel the mysteries of brain waves and cognition.

1.3   Aims and Scope of the Book

In this seminal exploration, "Quantum Mechanics of Brain Waves: A Cognitive Exploration," we embark on a captivating journey at the interface of quantum mechanics and neuroscience. This book aims to unravel the enigmatic relationship between brain waves and quantum phenomena, delving into the intriguing possibilities that arise from their interplay. We endeavor to present a comprehensive overview of the current state of research while delving into the potential implications and applications of understanding brain waves through the lens of quantum mechanics.

The primary aim of this book is to bridge the gap between two seemingly distinct domains - the macroscopic realm of cognitive neuroscience and the quantum world of subatomic particles. We seek to elucidate how principles from quantum mechanics might play a role in the complexities of brain function and cognition. By exploring various theoretical models, experimental evidence, and cutting-edge technologies, we aim to present a coherent narrative that fosters interdisciplinary collaboration and innovative thinking.

One of the core objectives is to introduce readers to the fundamentals of brain waves, including their classifications, characteristics, and relevance to cognitive processes. We provide a comprehensive overview of brain wave measurement techniques, such as electroencephalography (EEG), magnetoencephalography (MEG), and functional magnetic resonance imaging (fMRI), to understand how brain waves can be analyzed and correlated with cognitive states.

A central theme of the book is to explore the potential quantum nature of brain waves. We investigate theories proposing that brain waves might be influenced by quantum phenomena, such as quantum coherence, entanglement, and tunneling. We critically examine the Orchestrated Objective Reduction (Orch-OR) theory, which suggests that consciousness emerges from quantum computations in microtubules. Through a balanced evaluation of evidence and perspectives, we aim to foster informed discussions and inspire new avenues of research.

In addition to theoretical aspects, the book also delves into the exciting realm of experimental investigations. We explore recent studies that suggest the presence of quantum effects in biological systems and consider the implications for brain waves. Advanced techniques like quantum-enhanced imaging and quantum sensors offer promising opportunities to probe brain activity with unprecedented precision, potentially revealing subtle quantum phenomena.

Another significant aim of this book is to investigate the role of brain waves in various cognitive functions, including memory, perception, attention, and decision-making. We explore how understanding brain waves at the quantum level might shed light on the brain's computational capabilities and the mechanisms underlying cognitive processes.

Throughout the book, we provide real-world examples and case studies that exemplify the impact of brain waves on human behavior and cognition. We discuss how brain wave patterns might be related to neurological disorders, altered states of consciousness, and the potential for brain-computer interfaces that leverage quantum principles.

The ethical implications of understanding brain waves through quantum mechanics form an essential aspect of this exploration. We critically address questions related to cognitive enhancement, privacy concerns, and the responsible use of quantum technologies in the context of neuroscience.

The scope of this book extends beyond academic researchers and students. We aim to make the content accessible to a broader audience, including neuroscientists, physicists, psychologists, and anyone intrigued by the mysteries of the human mind. Through clear explanations, illustrative examples, and engaging discussions, we aspire to spark curiosity and inspire readers to contemplate the profound implications of quantum mechanics in the realm of brain waves and cognition.

In conclusion, "Quantum Mechanics of Brain Waves: A Cognitive Exploration" aims to unravel the fascinating relationship between quantum phenomena and brain function. By providing a comprehensive and accessible overview of brain waves and quantum mechanics, we aspire to foster a deeper appreciation for the complexity of the human brain and the captivating interplay between quantum mechanics and neuroscience.

Chapter 2

Foundations of Quantum Mechanics

2.1 Historical Development of Quantum Theory

The historical development of quantum theory marks a revolutionary era in physics that reshaped our understanding of the fundamental nature of reality. The journey began in the early 20th century, with physicists grappling with perplexing experimental observations that challenged classical physics. Through a series of groundbreaking discoveries and the collaborative efforts of brilliant minds, the quantum revolution unfolded, ushering in a new era of physics.

The story commences with Max Planck's seminal work on blackbody radiation in 1900. Planck introduced the concept of quantization of energy, suggesting that energy is emitted or absorbed in discrete units or "quanta." His revolutionary idea marked the birth of quantum theory, albeit unintentionally. Planck's work laid the foundation for understanding the behavior of light and matter at the atomic and subatomic levels.

Albert Einstein, building upon Planck's insights, made significant contributions to the understanding of the photoelectric effect in 1905. He postulated that light behaves as discrete packets of energy, now known as photons, which can eject electrons from a metal surface when absorbed. This work not only affirmed the quantized nature of energy but also played a crucial role in the development of quantum mechanics.

The Bohr model of the atom, proposed by Niels Bohr in 1913, introduced the idea of quantized electron orbits around the atomic nucleus. Bohr's model successfully explained the discrete spectral lines observed in atomic spectra, which could not be accounted for by classical physics. The quantization of energy levels in atoms was a pivotal step towards the quantum understanding of matter.

The real breakthrough came with the development of wave mechanics, initiated by Erwin Schrödinger and Werner Heisenberg independently in the mid-1920s. Schrödinger formulated the famous wave equation in 1926, describing the behavior of quantum systems in terms of wavefunctions:

$$i\hbar\frac{\partial\psi}{\partial t} = -\frac{\hbar^2}{2m}\nabla^2\psi + V\psi \tag{2.1}$$

where $\hbar$ is the reduced Planck's constant, m is the particle's mass, V is the potential energy, and ψ is the wavefunction.

Heisenberg, on the other hand, introduced matrix mechanics, where observables are represented by matrices. These two formulations were shown to be equivalent, and together they formed the foundation of quantum mechanics.

The principle of complementarity, proposed by Bohr, played a significant role in shaping the philosophical interpretation of quantum mechanics. Bohr emphasized that quantum entities can exhibit both particle-like and wave-like properties, depending on the experimental context. This idea challenged classical notions of objectivity and set the stage for the development of the Copenhagen interpretation of quantum mechanics.

The Copenhagen interpretation, championed by Bohr and Heisenberg, is a probabilistic interpretation of quantum mechanics. It asserts that until observed, quantum systems exist in a superposition of states, and the act of measurement causes the wavefunction to collapse to a specific state. This interpretation sparked debates about the role of the observer and the nature of reality in the quantum world.

The development of quantum mechanics was further enriched by the contributions of Paul Dirac, who introduced quantum field theory and contributed to the formulation of quantum electrodynamics. Dirac's work paved the way for the unification of quantum mechanics with special relativity, leading to profound insights into particle physics.

In 1927, the famous Solvay Conference brought together the leading physicists of the time, including Bohr, Einstein, Schrödinger, and Heisenberg, among others. The conference was a landmark event where the competing interpretations of quantum mechanics were intensely debated, leaving unresolved questions that continue to intrigue physicists to this day.

The development of quantum mechanics had a profound impact on various fields, including chemistry, solid-state physics, and astrophysics. It laid the groundwork for understanding the behavior of matter and energy at the atomic and subatomic scales, revolutionizing our comprehension of the natural world.

In conclusion, the historical development of quantum theory is a testament to human curiosity, ingenuity, and the quest for understanding the mysteries of the universe. From Planck's quantum hypothesis to the formulation of wave mechanics and the establishment of the Copenhagen interpretation, each milestone in this journey has shaped the landscape of modern physics. The profound implications of quantum mechanics extend far beyond physics, and this exploration of the historical roots of quantum theory sets the stage for delving into the interplay between quantum mechanics and brain waves in the subsequent chapters of the book.

2.2   Key Concepts: Superposition, Entanglement, and Uncertainty

In the captivating realm of quantum mechanics, three key concepts stand out as pillars of its mysterious foundation: superposition, entanglement, and uncertainty. These concepts challenge our classical intuitions and redefine the boundaries of reality at the quantum scale. In this section, we embark on an exploration of these fundamental principles, essential to grasp the essence of quantum mechanics and its potential implications for brain waves and cognition.

Superposition: The concept of superposition lies at the heart of quantum mechanics. It entails that quantum systems can exist in multiple states simultaneously, represented by their wavefunctions. Mathematically, if ψ_1 and ψ_2 are valid wavefunctions, then any linear combination of them, such as $c_1\psi_1 + c_2\psi_2$, where c_1 and c_2 are complex numbers, is also a valid wavefunction. This principle leads to phenomena such as interference patterns in quantum experiments. For example, in the double-slit experiment, particles can exhibit interference patterns when passing through two slits simultaneously, demonstrating their wave-like nature.

Entanglement: Entanglement is a striking phenomenon in quantum mechanics where the states of two or more particles become correlated in such a way that the statev of one particle cannot be described independently of the others. When particles become entangled, the quantum state of the entire system becomes a unified whole. Even if the entangled particles are spatially separated, their states remain interconnected, defying classical intuition. Entanglement has been experimentally verified and plays a crucial role in quantum information processing and quantum communication.

Uncertainty: The Heisenberg uncertainty principle is a fundamental aspect of quantum mechanics, expressing a limitation on the precision with which certain pairs of physical properties, such as position and momentum, can be simultaneously known. The principle states that the more precisely one property is measured, the less precisely the conjugate property can be known. For example, the position and momentum of a particle cannot be simultaneously measured with arbitrary accuracy. This principle introduces inherent indeterminacy at the quantum level and distinguishes quantum uncertainty from classical determinism.

Understanding these key concepts is essential for exploring the potential interplay between quantum mechanics and brain waves. The human brain is an intricately complex system of interconnected neurons, and its dynamics could be subject to quantum behavior. The delicate balance of superposition, entanglement, and uncertainty in quantum systems raises intriguing questions about whether such phenomena play a role in neural processing and cognition.

Several theoretical models propose that brain waves might arise from quantum processes within neural networks. The idea of quantum brain dynamics suggests that the brain's computational power might be enhanced through superposition and entanglement, enabling it to process information in unique and efficient ways.

One prominent hypothesis is that quantum coherence plays a role in brain function. The orchestrated objective reduction (Orch-OR) theory, proposed by Roger Penrose and Stuart Hameroff, suggests that consciousness arises from quantum computations in microtubules within neurons. While this theory remains controversial, it has ignited research into the potential connections between quantum mechanics and the mind.

Quantum brain dynamics could potentially explain puzzling aspects of consciousness and cognitive processes. The brain's ability to process vast amounts of information simultaneously might find its roots in the quantum phenomena of superposition and entanglement.

Experimental investigation is essential to unveil the quantum aspects of brain waves. Cutting-edge techniques such as quantum-enhanced imaging and quantum sensors offer promising avenues to probe brain activity with unprecedented precision. These advancements may uncover evidence of quantum coherence or entanglement in neural networks, providing invaluable insights into brain function.

The study of quantum mechanics and its potential implications for brain waves is not confined to neuroscience alone. It intersects with other disciplines, such as quantum biology and quantum information science. The interplay of quantum mechanics and brain waves transcends boundaries and fosters collaboration among physicists, neuroscientists, and cognitive scientists.

In conclusion, the key concepts of superposition, entanglement, and uncertainty constitute the bedrock of quantum mechanics, shaping our understanding of the quantum world. In the context of brain waves and cognition, these concepts open up exciting possibilities to explore the quantum nature of the mind. As we delve deeper into the mysteries of quantum mechanics and its potential connections to brain waves, we embark on a thrilling journey that unites quantum theory with the complexities of consciousness and the human brain.

2.3   Quantum States and Wavefunctions

In the wondrous world of quantum mechanics, the concept of quantum states and their mathematical description through wavefunctions forms the very essence of the theory. A quantum state represents the intrinsic properties of a quantum system, and the wavefunction encodes the probability amplitudes for each possible outcome of a measurement. In this section, we embark on a captivating journey to understand the intricacies of quantum states and wavefunctions, essential for comprehending the enigmatic behavior of the quantum realm.

A quantum state is a mathematical representation of a physical system in the language of quantum mechanics. In the context of quantum mechanics, a quantum system can exist in a superposition of states, meaning it can simultaneously be in multiple states at once. This property is fundamental to the behavior of particles at the quantum scale and leads to fascinating phenomena like interference.

The wavefunction is a central concept in quantum mechanics. It is a complex-valued function that describes the quantum state of a system as a function of its position or other relevant variables. For a single particle in one dimension, the wavefunction $\psi(x)$ provides the probability amplitude of finding the particle at position x. The probability of finding the particle within a certain range is given by $|\psi(x)|^2$, which is always a non-negative real number.

Mathematically, the wavefunction must satisfy normalization conditions to ensure the probabilities sum to unity over all possible positions. The normalization condition for a one-dimensional wavefunction is $\int_{-\infty}^{\infty} |\psi(x)|^2\, dx = 1$. This condition ensures that the total probability of finding the particle in the entire space is 100

Quantum states can also be represented as vectors in a mathematical space known as a Hilbert space. The wavefunction is a specific representation of the quantum state in a particular basis. The choice of basis affects how the wavefunction describes the system's behavior and how measurements are represented.

The time evolution of a quantum state is governed by the Schrödinger equation:

$$i\hbar \frac{\partial}{\partial t} \psi(x, t) = \hat{H} \psi(x, t) \tag{2.2}$$

where $\hbar$ is the reduced Planck's constant, $\psi(x, t)$ is the time-dependent wavefunction, and $\hat{H}$ is the Hamiltonian operator representing the total energy of the system.

The Schrödinger equation ensures that the wavefunction evolves smoothly and continuously with time. Solving the Schrödinger equation provides insight into how quantum systems change and how their properties may vary over time.

In quantum mechanics, measurements are represented by operators called observables. An observable $\hat{A}$ is represented by a Hermitian operator, and its corresponding eigenvalues and eigenvectors correspond to the possible outcomes and states of the measurement, respectively.

When a measurement is performed on a quantum system, the wavefunction collapses to one of the eigenstates of the measured observable. The probability of obtaining a specific eigenvalue is given by the square of the inner product between the wavefunction and the corresponding eigenstate.

The uncertainty principle, a fundamental tenet of quantum mechanics, places a limitation on the simultaneous measurability of certain pairs of observables. For example, the position and momentum of a particle cannot be precisely determined simultaneously, as expressed by the Heisenberg uncertainty principle:

$$\Delta x \cdot \Delta p \geq \frac{\hbar}{2} \tag{2.3}$$

where Δx is the uncertainty in position and Δp is the uncertainty in momentum.

Quantum states and wavefunctions play a vital role in understanding the behavior of particles and physical systems at the quantum scale. They are the foundation upon which the rich structure of quantum mechanics is built.

The concept of entanglement, where quantum systems become correlated in such a way that their states cannot be described independently, is intimately connected to quantum states and wavefunctions. Entangled states are described by composite wavefunctions that extend over multiple subsystems.

Entanglement has profound implications for quantum information processing and quantum communication. It enables quantum states to be used for tasks like quantum teleportation and quantum cryptography, which promise exciting advances in information technology.

The study of quantum states and wavefunctions is not limited to isolated particles; it extends to complex systems like atoms, molecules, and even macroscopic objects. The behavior of these systems is often described using methods like the density functional theory (DFT) for electronic structure calculations.

In conclusion, quantum states and wavefunctions lie at the core of quantum mechanics, defining the probabilities and behavior of quantum systems. Understanding these concepts is essential for comprehending the intricate nature of quantum phenomena. As we delve deeper into the mysteries of quantum states and their interplay with brain waves, we unlock the potential for groundbreaking insights into the relationship between quantum mechanics and the human mind.

2.4 The Measurement Problem and Collapse of the Wavefunction

The measurement problem is one of the most intriguing and debated aspects of quantum mechanics. It arises from the peculiar nature of quantum states and the process of measurement, leading to a fundamental question: What causes the transition from a superposition of states to a definite measurement outcome? This question has been the subject of intense scrutiny and various interpretations in quantum theory. In this section, we delve into the measurement problem and explore the concept of wavefunction collapse, shedding light on the challenges it poses to our understanding of the quantum world.

When a quantum system is in a superposition of states, it exists in a blend of all possible states with corresponding probability amplitudes. However, upon measurement, the system appears to "collapse" into a single definite state with a specific measurement outcome. This abrupt change from potentiality to actuality is at the core of the measurement problem.

One of the early and widely known interpretations addressing the measurement problem is the Copenhagen interpretation. Proposed by Niels Bohr and Werner Heisenberg, this interpretation advocates a probabilistic view of quantum measurements. According to the Copenhagen interpretation, the act of measurement causes the wavefunction to collapse into one of the possible states, with the probabilities of each outcome determined by the squared magnitudes of the probability amplitudes.

Another perspective on the measurement problem is the many-worlds interpretation, proposed by Hugh Everett in the 1950s. In this interpretation, the universe continually branches into multiple parallel realities, with each branch representing a different measurement outcome. This theory suggests that all possible measurement outcomes coexist in separate branches of the universe, preserving the unitary evolution of the wavefunction.

The de Broglie-Bohm interpretation, also known as pilot-wave theory, proposes an alternative approach to the measurement problem. In this interpretation, particles are guided by a hidden wave that influences their motion. The wavefunction remains intact, and particles follow deterministic trajectories, avoiding the need for wavefunction collapse.

Quantum decoherence is a concept that plays a significant role in addressing the measurement problem. It refers to the process by which a quantum system interacts with its environment, causing the superposition of states to become entangled with the environment. As a result, different branches of the wavefunction become isolated and effectively behave classically, leading to the appearance of wavefunction collapse without invoking the need for a conscious observer.

The role of the observer in the measurement process is a central point of contention in the measurement problem. Some interpretations, like the Copenhagen interpretation, emphasize the active role of the observer in causing the collapse of the wavefunction. Other interpretations, such as the many-worlds interpretation and decoherence theory, downplay the role of consciousness, attributing wavefunction collapse to interactions with the environment.

The measurement problem has profound implications for understanding the nature of reality and the role of consciousness in quantum mechanics. It challenges our classical intuitions and raises philosophical questions about the nature of observation and the relationship between the observer and the observed.

Resolving the measurement problem has been a central goal in the quest for a complete understanding of quantum mechanics. Various experiments and investigations continue to shed light on the dynamics of measurement and the behavior of quantum systems.

One of the remarkable applications of quantum mechanics lies in quantum computing. Quantum computers leverage the principles of superposition and entanglement to perform certain tasks exponentially faster than classical computers. The measurement problem and wavefunction collapse have significant consequences for quantum computing algorithms and their implementation.

The study of the measurement problem is not only of theoretical interest but also has practical implications for technology and information processing. Understanding the nature of measurement in quantum systems is essential for harnessing the power of quantum mechanics for real-world applications.

In conclusion, the measurement problem and wavefunction collapse are central mysteries of quantum mechanics, challenging our understanding of the quantum world. Various interpretations attempt to address this enigma, each offering unique perspectives on the nature of measurement and the behavior of quantum systems. As we explore the measurement problem and its implications, we venture into the frontier of quantum phenomena and the potential interplay with brain waves, enriching our understanding of the profound connections between quantum mechanics and cognition.

Chapter 3

The Neuroscience of Brain Waves

3.1 Brain Anatomy and Neuronal Communication

Understanding the brain's structure and the mechanisms of neuronal communication is fundamental to exploring the potential interplay between quantum mechanics and brain waves. The human brain, the most complex organ in the body, consists of billions of neurons interconnected through synapses, forming an intricate network that underlies all cognitive processes. In this section, we delve into the essential aspects of brain anatomy and the remarkable mechanisms by which neurons communicate, paving the way for comprehending the neural basis of quantum phenomena.

At the macroscopic level, the human brain can be divided into distinct regions, each associated with specific functions. The cerebral cortex, responsible for higher cognitive functions, memory, and perception, is a remarkable outer layer covering the cerebrum. Deep within the brain lies the thalamus, regulating sensory processing and relaying information to the cerebral cortex. The cerebellum, located at the back of the brain, plays a crucial role in motor control and coordination. Beneath the cerebrum and cerebellum is the brainstem, essential for vital functions like breathing and heart rate.

The neurons, the building blocks of the brain, are specialized cells that transmit and process information through electrical and chemical signals. Each neuron comprises a cell body, dendrites that receive signals from other neurons, and an axon that transmits signals to other neurons or cells. The synapse, a microscopic gap between neurons, facilitates communication through neurotransmitters.

Neuronal communication begins with the generation of electrical impulses, known as action potentials, at the neuron's cell body. When the neuron receives signals from other neurons through its dendrites, the membrane potential changes. If the potential reaches a certain threshold, an action potential is triggered, propagating along the axon.

The speed of action potential propagation is crucial for efficient communication. Neurons are wrapped in a fatty substance called myelin, which acts as an insulator, speeding up signal transmission. This myelination is essential for quick reflexes and coordinated movements.

At the synapse, the action potential triggers the release of neurotransmitters from the presynaptic neuron. These neurotransmitters traverse the synaptic gap and bind to receptors on the postsynaptic neuron, inducing electrical changes. Depending on the type of neurotransmitter and the receptor, the postsynaptic neuron may be excited or inhibited.

The integration of excitatory and inhibitory signals determines whether the postsynaptic neuron will fire an action potential. This process of synaptic transmission forms the basis of neural information processing and is fundamental to all brain functions.

The brain's complexity lies not only in its vast number of neurons but also in the intricate patterns of connectivity. Neurons form complex networks, enabling the brain to perform computations and process information in parallel.

The neural activity in the brain gives rise to various electrical patterns, including brain waves. These brain waves, measured using electroencephalography (EEG), represent synchronized electrical activity of a large group of neurons. Different brain wave frequencies are associated with specific states of consciousness, such as delta waves during deep sleep and gamma waves during intense cognitive processing.

The neural mechanisms underlying brain waves are of great interest to researchers exploring the potential connections between quantum mechanics and brain function. Some theoretical models propose that quantum processes within neurons might contribute to brain wave generation and cognitive functions. The interplay between quantum coherence and neural activity could potentially shape brain dynamics and cognition in unique ways.

Studying the intricate brain anatomy and neuronal communication is essential for unraveling the mysteries of brain waves and cognition. As we advance our understanding of these neural mechanisms, we open up exciting possibilities to explore the potential role of quantum mechanics in shaping brain waves and the complexities of the human mind.

3.2 Types of Brain Waves: Alpha, Beta, Theta, Delta, and Gamma

Brain waves, also known as neural oscillations, are rhythmic patterns of electrical activity that emerge from synchronized neural firing. These brain waves are detected through electroencephalography (EEG), a non-invasive technique that records electrical signals from the scalp. The study of brain waves has been integral to understanding brain function and cognitive processes. In this section, we explore five major types of brain waves - Alpha, Beta, Theta, Delta, and Gamma - each associated with distinct cognitive states and physiological activities.

1. **Alpha Waves**: Alpha waves are prominent in the brain when a person is awake but relaxed with their eyes closed. They have a frequency range of

approximately 8 to 13 Hz and are typically observed in the occipital lobe. Alpha waves play a crucial role in facilitating mental relaxation and reducing sensory processing. They are also associated with creative insights and a state of calm focus.

2. **Beta Waves**: Beta waves have a higher frequency range of around 14 to 30 Hz and are prevalent when the brain is in an alert and active state. They are most dominant during waking hours and are prominent in the frontal cortex. Beta waves are linked to focused attention, problem-solving, and active cognitive processing.

3. **Theta Waves**: Theta waves have a frequency range of approximately 4 to 7 Hz and are observed during drowsiness, light sleep, and deep meditation. They are prominently recorded in the hippocampus and are associated with memory consolidation, learning, and deep relaxation.

4. **Delta Waves**: Delta waves have the slowest frequency range, typically below 4 Hz. They are most prevalent during deep sleep and are crucial for restorative and regenerative processes in the brain and body. Delta waves are associated with dreamless sleep and the release of growth hormones.

5. **Gamma Waves**: Gamma waves have the highest frequency, ranging from 30 to 100 Hz or even higher. They are observed during intense cognitive processing, problem-solving, and heightened states of consciousness. Gamma waves are believed to play a role in binding together different brain regions to facilitate complex cognitive functions.

The precise mechanisms underlying brain wave generation are still a subject of ongoing research. However, it is clear that brain waves result from the collective activity of large groups of neurons firing in synchrony.

One intriguing aspect of brain waves is their potential connection with quantum mechanics. Some theoretical models propose that the dynamics of brain waves could be influenced by quantum phenomena, such as quantum coherence and entanglement within neural networks. Understanding the interplay between quantum mechanics and brain waves may offer novel insights into the neural basis of consciousness and cognition.

In various states of consciousness, brain waves can exhibit dynamic changes. For example, during focused attention, beta waves may dominate, while alpha waves may increase during states of relaxation and meditation. Sleep stages are characterized by shifts in brain wave patterns, with delta waves predominating in deep sleep and theta waves prominent during rapid eye movement (REM) sleep, associated with dreaming.

The study of brain waves has practical applications in diverse fields. Neurofeedback, a technique that enables individuals to self-regulate their brain activity, is used for therapeutic purposes in treating conditions like anxiety, attention deficit hyperactivity disorder (ADHD), and post-traumatic stress disorder (PTSD).

Researchers are also exploring the potential use of brain waves in brain-computer interfaces (BCIs) for controlling external devices or aiding individuals with motor impairments. Brain waves have also been investigated in fields like sleep research, meditation, and cognitive enhancement.

While the role of quantum mechanics in brain waves remains an area of speculation, the study of brain wave patterns and their implications for cognition is an exciting avenue for interdisciplinary research. As we uncover more about the intricate world of brain waves, we gain valuable insights into the complexities of human consciousness and the potential interplay with quantum mechanics.

3.3 Brain Wave Measurement Techniques: EEG, MEG, and fMRI

The study of brain waves has been revolutionized by advanced brain imaging techniques that enable non-invasive measurements of neural activity. Three prominent methods used to investigate brain waves are Electroencephalography (EEG), Magnetoencephalography (MEG), and functional Magnetic Resonance Imaging (fMRI). Each technique offers unique advantages and provides valuable insights into the dynamic functioning of the human brain. In this section, we explore the principles, applications, and limitations of these brain wave measurement techniques.

1. **Electroencephalography (EEG)**: EEG is a widely used brain imaging technique that records electrical activity from the scalp. It measures brain waves resulting from the synchronized activity of large groups of neurons. EEG is particularly valuable in studying the temporal dynamics of brain waves with high temporal resolution, on the order of milliseconds. It is well-suited for examining cognitive processes, sleep stages, and monitoring brain activity in real-time during various tasks. However, EEG has limited spatial resolution due to the volume conduction effect, where electrical signals from deep brain structures may appear diffuse on the scalp.

2. **Magnetoencephalography (MEG)**: MEG measures the magnetic fields generated by neuronal electrical activity using superconducting sensors. Like EEG, MEG provides high temporal resolution but offers better spatial resolution. MEG is particularly useful for localizing brain activity and identifying the sources of brain waves. However, MEG systems are expensive and require specialized facilities due to the need for cryogenic cooling of the superconducting sensors.

3. **Functional Magnetic Resonance Imaging (fMRI)**: fMRI measures changes in blood flow and oxygenation in the brain, providing an indirect measure of neural activity. It offers excellent spatial resolution, allowing researchers to map brain activation to specific brain regions. fMRI is widely used in cognitive neuroscience and has helped uncover brain networks associated with various cognitive functions. However, fMRI has lower temporal resolution compared to EEG and MEG, typically on the order of seconds.

Combining these brain wave measurement techniques allows researchers to obtain complementary information about brain activity. For example, simultaneous EEG-fMRI recordings enable the correlation of EEG-derived brain waves with fMRI-based brain activation, providing a more comprehensive understanding of neural processing.

The applications of these brain wave measurement techniques are vast. EEG is commonly used in clinical settings for diagnosing neurological disorders such as epilepsy and sleep disorders. It is also utilized in brain-computer interfaces (BCIs) for controlling external devices using brain waves. MEG is valuable in localizing epileptic foci before surgery and investigating brain activity in language and sensory processing.

fMRI has significantly contributed to the study of brain function in cognitive tasks, emotion, memory, and decision-making. It is a valuable tool in research related to brain disorders, such as Alzheimer's disease and schizophrenia.

In recent years, researchers have explored the potential interplay between brain waves and quantum phenomena. Some theoretical models propose that quantum coherence in neural networks could influence brain wave patterns and cognition. Integrating brain wave measurements with quantum-sensitive techniques may pave the way for investigating the role of quantum mechanics in brain function and consciousness.

However, there are challenges in interpreting brain wave measurements. Brain waves are complex, and their relationships with specific cognitive processes are still being unraveled. Additionally, variations in brain wave patterns among individuals make it essential to consider personalized approaches in analyzing brain wave data.

In conclusion, brain wave measurement techniques, including EEG, MEG, and fMRI, have revolutionized the study of brain activity and cognitive processes. These techniques offer complementary information, enabling researchers to gain a deeper understanding of brain function. Advancements in brain wave measurement and quantum-sensitive techniques may uncover novel insights into the neural basis of cognition and the potential connections between brain waves and quantum mechanics.

3.4 Brain Waves and States of Consciousness

The study of brain waves provides valuable insights into the intricate relationship between neural activity and states of consciousness. Different types of brain waves are associated with distinct cognitive and physiological states, ranging from wakefulness to deep sleep. In this section, we explore how brain waves reflect different states of consciousness and how alterations in brain wave patterns can have profound effects on our perception, attention, and awareness.

1. **Wakefulness and Alpha Waves**: During wakefulness with closed eyes, alpha waves dominate the brain. These 8 to 13 Hz oscillations are most prominent in the occipital lobe and are associated with relaxed alertness. When individuals close their eyes and enter a state of relaxation, alpha waves become more prominent, reflecting a calm and focused mind.

2. **Focused Attention and Beta Waves**: As we engage in mental tasks requiring focused attention, beta waves take over. These higher-frequency brain waves (14 to 30 Hz) are prevalent in the frontal cortex and reflect heightened cognitive processing. Beta waves help us concentrate on specific stimuli and perform tasks requiring sustained mental effort.

3. **Meditation and Theta Waves**: During meditation or deep relaxation, theta waves (4 to 7 Hz) become more pronounced. These brain waves are prominent in the hippocampus and are associated with introspection, creativity, and a deeply relaxed state of mind.

4. **Deep Sleep and Delta Waves**: When we enter deep sleep, our brain wave activity slows down dramatically, and delta waves (below 4 Hz) dominate. These slow, high-amplitude waves are crucial for restorative sleep, memory consolidation, and overall well-being.

5. **REM Sleep and Dreaming**: Rapid Eye Movement (REM) sleep is characterized by a mix of brain wave activity, including beta and alpha waves, resembling wakefulness. During this stage, our most vivid dreams occur, reflecting a state of heightened brain activity despite the body's muscle paralysis.

Altering brain wave patterns can lead to various changes in consciousness. For example, meditation practices that promote theta wave activity are associated with a sense of inner calm and enhanced introspection. On the other hand, sleep disorders that disrupt normal brain wave patterns can lead to problems with memory, attention, and overall cognitive function.

Neurological conditions like epilepsy can also result from abnormal brain wave activity. During an epileptic seizure, there is a sudden and intense synchronization of neuronal firing, leading to abnormal brain waves and altered consciousness.

Furthermore, brain wave entrainment is a phenomenon where external stimuli, such as rhythmic sounds or visual patterns, synchronize brain waves to the same frequency. This technique has been explored for its potential to induce altered states of consciousness, relaxation, and even pain reduction.

The study of brain waves and their correlation with consciousness is an essential aspect of neuroscience. Advances in brain wave measurement techniques, such as EEG and MEG, have enabled researchers to investigate brain dynamics in real-time during various cognitive tasks and states of consciousness.

While the role of quantum mechanics in consciousness remains a subject of debate, some theoretical models suggest that quantum coherence within neural networks could contribute to our experience of consciousness. This notion opens up exciting avenues for interdisciplinary research at the intersection of quantum mechanics and neuroscience.

In conclusion, brain waves play a crucial role in shaping states of consciousness, ranging from focused attention and wakefulness to deep sleep and meditation. The interplay between brain wave patterns and cognitive states is of immense significance for understanding human perception, awareness, and neural processing. As we delve deeper into the complexities of brain waves and their potential connections to quantum mechanics, we unlock new insights into the nature of consciousness and the profound mysteries of the human mind.

Chapter 4

The Quantum Nature of Neurons

4.1 Quantum Mechanics at the Cellular Level

The classical view of biology considers living organisms as deterministic systems governed by classical physics. However, at the cellular level, especially in the realm of neurons, emerging evidence suggests that quantum mechanics might play a significant role. The study of quantum phenomena at the cellular level is a fascinating area of research, exploring the potential interplay between quantum mechanics and biological processes. In this section, we delve into the intriguing world of quantum mechanics in neurons, examining the key aspects that make this realm unique.

One essential concept in quantum mechanics is *superposition*, where quantum systems can exist in multiple states simultaneously until observed. This notion challenges our classical intuition, but it has been experimentally demonstrated in isolated quantum systems. In neurons, some researchers propose that quantum superposition might occur within proteins, ions, or even microtubules, fundamental structures within neurons. The idea is that quantum superposition could potentially influence neural firing and information processing, leading to novel insights into cognitive functions.

Another intriguing aspect is *quantum entanglement*, where the states of two or more particles become correlated in such a way that the state of one particle instantly affects the state of another, regardless of distance. While entanglement is typically associated with particles at the atomic or subatomic level, researchers have speculated that entangled states might extend to larger scales within neurons, possibly influencing the coherence and dynamics of neural networks.

The *quantum tunneling* phenomenon, where particles can pass through energy barriers that would be classically insurmountable, is also relevant to cellular quantum processes. Some studies suggest that quantum tunneling might facilitate energy transfer within biological systems, such as the efficient transfer of electrons during photosynthesis. While the extent of quantum tunneling in neurons is still an open question, it offers exciting possibilities for understanding the efficiency of energy transfer within the brain.

Quantum mechanics introduces the concept of *quantum coherence*, where quantum systems can maintain phase relationships and exhibit wave-like properties. Quantum coherence is highly sensitive to the surrounding environment, and any interaction with the environment can lead to *quantum decoherence*, where the quantum nature of the system is lost. Researchers investigating quantum effects in neurons aim to understand the delicate balance between quantum coherence and decoherence, and how it might relate to information processing and cognitive functions.

The *Heisenberg uncertainty principle* is a fundamental concept in quantum mechanics, stating that certain pairs of physical properties, such as position and momentum, cannot be precisely measured simultaneously. While this principle is often associated with microscopic particles, some researchers have speculated that uncertainty at the cellular level might have functional implications for signal processing in neurons.

One intriguing hypothesis is the *Orch OR theory*, proposed by Roger Penrose and Stuart Hameroff, which suggests that quantum computations within microtubules could contribute to consciousness. According to this theory, microtubules act as quantum computers, and their interactions give rise to coherent brain-wide quantum states that underlie conscious experience. The Orch OR theory has generated significant debate, and its experimental validation remains a topic of active research.

Despite the exciting possibilities, studying quantum mechanics at the cellular level is challenging. The delicate nature of quantum systems makes them highly susceptible to environmental noise and decoherence. Additionally, precise measurements of quantum effects in neurons present technical difficulties, as they require advanced tools and methodologies.

Advances in experimental techniques, such as quantum microscopy and quantum sensors, are paving the way for more refined investigations into the quantum nature of neurons. Researchers are also exploring the potential role of quantum coherence in neural computation and the potential connections between quantum phenomena and neural information processing.

Understanding the quantum aspects of neurons is not only a scientific curiosity but also holds significant implications for brain function and cognition. Exploring the potential interplay between quantum mechanics and cellular processes may shed light on the nature of consciousness, memory, and learning, unraveling the mysteries of the human brain and its cognitive capabilities.

In conclusion, the study of quantum mechanics at the cellular level, particularly within neurons, is a fascinating and rapidly evolving field of research. Concepts like superposition, entanglement, tunneling, and coherence open up exciting possibilities for understanding the quantum nature of biological systems. While the role of quantum mechanics in neurons and cognitive functions is still a topic of exploration and debate, the potential connections between quantum phenomena and brain activity present a promising frontier in neuroscience. As we continue to unravel the quantum mysteries of neurons, we embark on a journey that may revolutionize our understanding of brain waves, consciousness, and the complexities of the human mind.

4.2   Ion Channels and Quantum Tunneling

In neurons, the transmission of electrical signals relies on the opening and closing of ion channels embedded in the cell membrane. These ion channels play a critical role in regulating the flow of ions, such as sodium, potassium, and calcium, across the cell membrane, leading to the generation and propagation of electrical impulses. While classical models satisfactorily describe ion channel behavior, recent studies have begun to explore the potential influence of quantum tunneling on ion channel dynamics.

Quantum tunneling is a remarkable phenomenon in quantum mechanics, where a particle can pass through a potential energy barrier despite lacking sufficient classical energy to surmount it. Instead of overcoming the barrier classically, the particle exhibits wave-like properties and can appear on the other side of the barrier with a certain probability. In the context of ion channels, quantum tunneling could allow ions to cross the cell membrane more efficiently, enhancing the overall transmission of electrical signals.

One of the key ion channels implicated in quantum tunneling is the *voltage-gated potassium (K+) channel*. This channel is essential for regulating the electrical excitability of neurons. Quantum tunneling through this channel may enable potassium ions to pass through the channel more readily, leading to a faster repolarization phase of the action potential.

Quantum tunneling in ion channels raises intriguing questions about the role of quantum phenomena in neural information processing. The speed and efficiency of signal transmission in neurons are crucial for cognitive functions and decision-making processes. If quantum tunneling indeed plays a significant role in ion channel dynamics, it could influence the overall speed and accuracy of neural signaling.

One challenge in studying quantum tunneling in ion channels is the delicate nature of quantum systems and their susceptibility to environmental factors. Ion channels are embedded in a complex cellular environment, and the presence of surrounding molecules could impact quantum effects. Researchers are actively investigating how factors like temperature, ion concentrations, and membrane potential influence quantum tunneling within ion channels.

Moreover, the exploration of quantum tunneling in ion channels requires advanced experimental techniques and theoretical models. Techniques like electrophysiology and quantum microscopy are essential for studying ion channel behavior at the quantum level. Additionally, quantum mechanical simulations and computational models are used to understand the probability of quantum tunneling events in ion channels.

Understanding the interplay between ion channels and quantum tunneling has the potential to shed light on the fundamental principles of neural communication and computation. It may reveal how the brain efficiently processes and transmits information while maintaining a balance between classical and quantum behaviors.

Furthermore, investigations into quantum tunneling in ion channels can lead to new strategies for drug development. Many medications target ion channels to modulate neural activity. If quantum effects are found to significantly influence ion channel behavior, novel drugs could be designed to exploit or control quantum tunneling for therapeutic purposes.

As we delve deeper into the quantum nature of neurons, the potential implications of quantum tunneling in ion channels extend beyond neuroscience. Quantum tunneling is a phenomenon that extends to various fields of physics, including atomic and molecular systems. The study of ion channels could provide a unique window into understanding quantum phenomena in a biological context.

The complexity of ion channel dynamics and their potential quantum behaviors make this area of research a rich and promising frontier. As technologies advance and our understanding of quantum mechanics grows, we are poised to unravel the intricate quantum nature of neurons and its implications for brain function, cognition, and the very essence of consciousness.

In conclusion, ion channels play a crucial role in neural communication, and recent studies suggest that quantum tunneling may influence their dynamics.

The phenomenon of quantum tunneling, with its wave-like properties and probabilistic nature, could lead to faster and more efficient ion flow through channels. Understanding the role of quantum tunneling in ion channels is a challenging yet exciting endeavor with profound implications for neuroscience, drug development, and our understanding of quantum phenomena in biological systems. As we unlock the mysteries of ion channel dynamics at the quantum level, we gain deeper insights into the quantum nature of neurons and its potential impact on brain waves, neural information processing, and the complexities of the human mind.

Quantum entanglement is another fundamental concept in quantum information processing. When qubits become entangled, the state of one qubit instantaneously influences the state of another, regardless of distance. This property can enable instantaneous communication and coordination of information within neural networks. If quantum entanglement exists in neurons, it could underpin rapid and synchronized firing patterns observed during cognitive tasks.

One exciting possibility in quantum information processing is *quantum parallelism*. Quantum computers can perform certain calculations exponentially faster than classical computers by harnessing the power of superposition and entanglement. In the context of neurons, this could lead to more efficient processing of complex information and potentially explain the brain's remarkable computational capabilities.

However, the quantum nature of information processing in neurons also poses challenges. *Quantum decoherence*, caused by interactions with the environment, can disrupt quantum states and lead to loss of information. The brain's warm and noisy environment could significantly impact the delicate quantum processes, limiting the duration and extent of quantum information processing in neurons.

The Orch OR theory, proposed by Roger Penrose and Stuart Hameroff, suggests that quantum computations within microtubules contribute to consciousness. According to this theory, quantum coherence in microtubules can lead to brain-wide coherent states that generate conscious experience. While the

Orch OR theory remains controversial, it highlights the potential significance of quantum information processing in the brain.

Quantum information processing in neurons could also provide a novel perspective on brain wave patterns. Brain waves reflect the collective electrical activity of neural networks. If quantum entanglement or parallelism occurs within neurons, it could result in coordinated brain wave patterns, enhancing information transfer and synchronization between brain regions.

The study of quantum information processing in neurons requires interdisciplinary collaboration between quantum physicists and neuroscientists. Techniques for probing quantum phenomena, such as quantum microscopy, must be integrated with neuroscience methodologies like electrophysiology to explore potential quantum effects in neural circuits.

Moreover, experimental verification of quantum information processing in neurons is challenging due to the fragility of quantum states. Advanced techniques, including quantum error correction and quantum control, might be necessary to manipulate and measure quantum processes in living neurons.

Furthermore, the potential implications of quantum information processing in neurons extend beyond neuroscience. Quantum information science is a rapidly developing field with applications in cryptography, communication, and computation. The insights gained from studying quantum information processing in neurons could inspire new approaches to quantum computing and communication technologies.

Quantum mechanics at the cellular level, especially within neurons, presents an exciting frontier for research. The possibility of quantum information processing in the brain challenges traditional views of neural computation and cognition. By exploring the quantum nature of neurons, we gain deeper insights into the complexities of brain waves, information processing, and the enigmatic nature of consciousness.

In conclusion, quantum information processing in neurons is an emerging field with profound implications for neuroscience and quantum physics. Concepts like quantum superposition, entanglement, and parallelism open up exciting possibilities for understanding the brain's computational power and synchronization of neural activity. While challenges exist in verifying quantum effects in living neurons, the potential breakthroughs in quantum information science could revolutionize our understanding of brain waves, cognitive functions, and the nature of consciousness.

Chapter 5

Quantum Mechanics and Cognitive Processes

5.1 Quantum Information Processing in Neurons

The brain is often referred to as the most sophisticated information processing system known. Traditional neuroscience attributes neural information processing to classical computations involving electrical signals and chemical neurotransmitters. However, the emerging field of *quantum information processing* explores the possibility that quantum phenomena within neurons could contribute to novel and efficient forms of information processing. In this section, we delve into the potential role of quantum information processing in neurons and its implications for brain waves and cognitive functions.

One key concept in quantum information processing is *quantum superposition*. Unlike classical bits, which can be either 0 or 1, quantum bits or *qubits* can exist in a superposition of both 0 and 1 states simultaneously. This property allows qubits to process multiple pieces of information in parallel. In neurons, it is hypothesized that certain biomolecules, such as ions or proteins, could act as qubits, enabling simultaneous processing of multiple neural signals.

Quantum entanglement is another fundamental concept in quantum information processing. When qubits become entangled, the state of one qubit instantaneously influences the state of another, regardless of distance. This property can enable instantaneous communication and coordination of information within neural networks. If quantum entanglement exists in neurons, it could underpin rapid and synchronized firing patterns observed during cognitive tasks.

One exciting possibility in quantum information processing is *quantum parallelism*. Quantum computers can perform certain calculations exponentially faster than classical computers by harnessing the power of superposition and entanglement. In the context of neurons, this could lead to more efficient processing of complex information and potentially explain the brain's remarkable computational capabilities.

However, the quantum nature of information processing in neurons also poses challenges. *Quantum decoherence*, caused by interactions with the environment, can disrupt quantum states and lead to loss of information. The brain's warm and noisy environment could significantly impact the delicate quantum processes, limiting the duration and extent of quantum information processing in neurons.

The Orch OR theory, proposed by Roger Penrose and Stuart Hameroff, suggests that quantum computations within microtubules contribute to consciousness. According to this theory, quantum coherence in microtubules can lead to brain-wide coherent states that generate conscious experience. While the Orch OR theory remains controversial, it highlights the potential significance of quantum information processing in the brain.

Quantum information processing in neurons could also provide a novel perspective on brain wave patterns. Brain waves reflect the collective electrical activity of neural networks. If quantum entanglement or parallelism occurs within neurons, it could result in coordinated brain wave patterns, enhancing information transfer and synchronization between brain regions.

The study of quantum information processing in neurons requires interdisciplinary collaboration between quantum physicists and neuroscientists. Techniques for probing quantum phenomena, such as quantum microscopy, must be integrated with neuroscience methodologies like electrophysiology to explore potential quantum effects in neural circuits.

Moreover, experimental verification of quantum information processing in neurons is challenging due to the fragility of quantum states. Advanced techniques, including quantum error correction and quantum control, might be necessary to manipulate and measure quantum processes in living neurons.

Furthermore, the potential implications of quantum information processing in neurons extend beyond neuroscience. Quantum information science is a rapidly developing field with applications in cryptography, communication, and computation. The insights gained from studying quantum information processing in neurons could inspire new approaches to quantum computing and communication technologies.

Quantum mechanics at the cellular level, especially within neurons, presents an exciting frontier for research. The possibility of quantum information processing in the brain challenges traditional views of neural computation and cognition. By exploring the quantum nature of neurons, we gain deeper insights into the complexities of brain waves, information processing, and the enigmatic nature of consciousness.

In conclusion, quantum information processing in neurons is an emerging field with profound implications for neuroscience and quantum physics. Concepts like quantum superposition, entanglement, and parallelism open up exciting possibilities for understanding the brain's computational power and synchronization of neural activity. While challenges exist in verifying quantum effects in living neurons, the potential breakthroughs in quantum information science could revolutionize our understanding of brain waves, cognitive functions, and the nature of consciousness.

5.2 Decision-Making and Superposition of States

Decision-making is a fundamental cognitive process that plays a crucial role in human behavior. Traditional models of decision-making are based on classical physics, assuming that individuals assess all possible options and choose the one with the highest utility. However, the complexity of decision-making often leads to cognitive biases and irrational behaviors. The emerging field of *quantum cognition* proposes an alternative approach, where decision-making is modeled using principles from quantum mechanics. In this section, we explore how superposition of states in quantum cognition can provide new insights into decision-making processes.

In quantum mechanics, a *superposition* is a state where a particle or a system exists in multiple states simultaneously, with each state represented by a probability amplitude. When a measurement is made, the system collapses into one of these states with a certain probability. In quantum cognition, this concept is applied to represent how individuals entertain multiple options or beliefs simultaneously during decision-making.

An example of superposition in decision-making is the well-known *Ellsberg paradox*, where individuals prefer an outcome with a known probability over an outcome with an unknown probability, even when the expected value suggests otherwise. Quantum cognition models explain this paradox by considering that individuals might superpose their beliefs about the unknown probability with the known probabilities, leading to the preference for a known outcome.

Another intriguing concept in quantum cognition is *quantum interference*, which occurs when the probability amplitudes of different states interfere constructively or destructively. Quantum interference can lead to interesting decision-making phenomena, such as the *order effect*, where the sequence of presenting choices influences decision preferences. Quantum models suggest that the interference between options presented earlier and later can result in non-linear decision patterns.

Furthermore, *entanglement* in quantum cognition refers to the interdependence of decision contexts and the entangled nature of beliefs about different aspects of a decision. For example, in medical decision-making, a patient's perception of treatment risks and benefits might be entangled with their emotional state and past experiences. Quantum models propose that such entanglement could better capture the complex interplay of factors influencing decisions.

Quantum cognition provides a fresh perspective on the *duality of thinking*, where individuals can simultaneously hold conflicting beliefs or preferences. Classical models struggle to explain this phenomenon, but quantum superposition elegantly addresses it. By allowing for the coexistence of contradictory mental states, quantum cognition offers a potential explanation for ambivalent decision-making.

One intriguing hypothesis is that the brain might exploit quantum computation for decision-making processes. Neural structures like *microtubules* within neurons have been proposed as possible candidates for quantum processing. The Orch OR theory suggests that microtubules' quantum properties contribute to consciousness and cognitive functions, including decision-making.

While quantum cognition presents exciting possibilities for understanding decision-making, it also raises questions about the neural mechanisms that might support quantum-like processing in the brain. Identifying neural structures and processes that can exhibit quantum behavior is a challenging task that requires interdisciplinary collaboration between quantum physicists, neuroscientists, and psychologists.

Moreover, experimental validation of quantum cognition models is essential to determine their explanatory power and predictive accuracy. Behavioral experiments that test decision-making under different conditions can provide empirical evidence for or against the quantum nature of cognitive processes.

The integration of quantum mechanics and cognitive processes is a young and dynamic field. It holds the promise of addressing long-standing puzzles in decision-making, such as irrational biases and preferences, and can potentially lead to more accurate and nuanced models of human behavior.

In conclusion, the intersection of quantum mechanics and cognitive processes has given rise to the field of quantum cognition, which provides a fresh perspective on decision-making. Superposition of states, quantum interference, entanglement, and the duality of thinking offer new insights into the complexities of human decision-making. While quantum cognition models are still in their early stages and face empirical and theoretical challenges, they open up exciting possibilities for a deeper understanding of the brain's cognitive capabilities and the enigmatic nature of human decisions.

Perception and attention are fundamental cognitive processes that allow us to make sense of the world around us. Traditional cognitive psychology explains these processes using classical models, but recent research in the field of *quantum cognition* suggests that quantum mechanics might offer a new perspective. In this section, we explore the concept of *quantum interference* in perception and attention and how it could shape our understanding of these cognitive phenomena.

In quantum mechanics, interference occurs when two or more quantum states overlap and interact. This leads to constructive or destructive interference, where the probabilities of certain outcomes are enhanced or suppressed. In the context of perception, quantum interference could explain how our brain processes and integrates sensory information from different modalities, such as vision, hearing, and touch.

For example, the McGurk effect is a perceptual phenomenon where conflicting auditory and visual inputs lead to a fused, perceived sound. Quantum cognition models propose that the interference between auditory and visual states might underpin this phenomenon. The brain's ability to integrate sensory inputs in a context-dependent manner could be a result of quantum interference effects.

Moreover, quantum interference might also play a role in attentional processes. Attention allows us to focus on relevant stimuli while filtering out distractions. Quantum models suggest that attentional states could exist in superposition, where multiple possible attentional focuses coexist simultaneously. This could explain how attention can shift rapidly and flexibly between different aspects of a stimulus.

An intriguing aspect of quantum interference in perception and attention is its potential relevance to *perceptual binding*. Perceptual binding refers to the brain's ability to integrate features of a stimulus into a coherent whole. Quantum interference could explain how different features of a stimulus become entangled and processed as a unified percept.

Furthermore, quantum interference in perception and attention may have implications for understanding *visual illusions*. Illusions occur when our perception deviates from objective reality. For instance, the Necker cube illusion presents an ambiguous 3D structure that appears to switch orientations. Quantum interference models propose that perceptual dynamics, including the alternation of interpretations, could arise from the superposition of multiple perceptual states.

An interesting aspect of quantum interference in perception and attention is its potential connection to *consciousness*. Some theories suggest that consciousness emerges from the integration and selection of information in the brain. Quantum interference effects might play a role in the brain's capacity to integrate and process information in a non-classical manner, potentially contributing to the rich subjective experience of consciousness.

Quantum interference in perception and attention also raises questions about the neural mechanisms that might underlie these effects. Identifying neural structures and processes that can exhibit quantum behavior is a complex challenge that requires collaboration between quantum physicists, neuroscientists, and psychologists.

Experimental investigations are essential for validating quantum interference models in perception and attention. Behavioral experiments that explore interference effects in perception and attention could provide empirical evidence for or against the quantum nature of cognitive processes.

The integration of quantum mechanics and cognitive processes is a fascinating area of research. It challenges traditional views of perception and attention, offering new insights into the brain's ability to process and interpret sensory information. Quantum interference has the potential to enrich our understanding of how we perceive the world and direct our attention in complex and dynamic environments.

In conclusion, quantum interference in perception and attention is an emerging concept that could revolutionize our understanding of cognitive processes. Superposition of perceptual states and the entanglement of attentional focuses open up new possibilities for explaining phenomena such as the McGurk effect, perceptual binding, and visual illusions. While the field of quantum cognition is still in its infancy, it holds the promise of providing deeper insights into the complexities of perception, attention, and consciousness.

5.3   Quantum Computation Models of Cognition

The intersection of quantum mechanics and cognitive processes has led to the development of an intriguing field known as *quantum cognition*. A fascinating aspect of this field is the exploration of *quantum computation models* of cognition. These models propose that certain cognitive processes might be more efficiently explained and simulated using principles from quantum computation. In this section, we delve into the concept of quantum computation models of cognition and their potential implications for our understanding of the human mind.

At the core of quantum computation models lie *quantum algorithms*, which harness the unique properties of superposition and entanglement to perform computations. One well-known quantum algorithm is *Grover's algorithm*, which can search an unsorted database in quadratically fewer steps than classical algorithms.

Quantum computation models suggest that the brain might leverage quantum algorithms to perform cognitive tasks more efficiently. For example, memory retrieval, a fundamental cognitive process, could benefit from *quantum parallelism*. Instead of sequentially searching through all stored memories, the brain could simultaneously explore multiple memory states through quantum superposition, leading to faster and more accurate retrieval.

Another intriguing cognitive process that quantum computation models address is *associative learning*. Classical models often struggle to explain how the brain efficiently forms and retrieves associations between different concepts. Quantum algorithms, such as *amplitude amplification*, provide a potential mechanism for rapid association formation and retrieval through constructive interference.

Moreover, quantum computation models shed light on *creativity and insight*. The process of generating creative ideas involves exploring a vast solution space to find novel and valuable solutions. Quantum algorithms like *quantum walks* and *quantum annealing* offer more efficient ways for the brain to explore and optimize this solution space, leading to moments of insight.

Quantum computation models also propose explanations for *decision-making under uncertainty*. Making decisions in uncertain or ambiguous situations is a challenging cognitive task. Classical models often struggle to account for the complexities of such decisions. Quantum algorithms, like *quantum Bayesian networks*, offer a promising approach to model decision-making under uncertainty by considering superposed beliefs about different possible outcomes.

Furthermore, *quantum machine learning* provides a framework for understanding how the brain could learn and adapt based on experience. Quantum machine learning algorithms, such as *quantum support vector machines* and *quantum reinforcement learning*, might offer more efficient ways for the brain to process and generalize information from the environment.

Despite the promise of quantum computation models of cognition, they face significant challenges. One critical challenge is the *decoherence problem*. Quantum systems are highly sensitive to interactions with the environment, leading to the loss of quantum coherence and the emergence of classical behavior. Maintaining quantum coherence in the brain's warm and noisy environment is a major hurdle for the feasibility of quantum computation models.

Another challenge is *scalability*. Quantum algorithms often require a large number of qubits to achieve their speedup advantages. The human brain is composed of billions of interconnected neurons, and emulating quantum computation at such a scale is currently beyond our technological capabilities.

Experimental validation of quantum computation models is essential for their acceptance and refinement. Brain imaging techniques, such as *functional magnetic resonance imaging (fMRI)* and *electroencephalography (EEG)*, could be used to investigate whether brain activity exhibits quantum behavior during specific cognitive tasks.

Despite these challenges, quantum computation models of cognition represent a cutting-edge field that offers new perspectives on the computational power of the brain. By integrating quantum mechanics with cognitive processes, we gain deeper insights into how the mind operates and addresses complex tasks.

In conclusion, quantum computation models of cognition are an exciting frontier of research that explores the potential benefits of quantum algorithms for explaining and simulating cognitive processes. From memory retrieval to decision-making and creativity, quantum algorithms offer new possibilities for understanding the mind's computational capabilities. However, challenges like decoherence and scalability must be addressed for these models to become fully viable. As technology advances and our understanding of quantum cognition grows, quantum computation models have the potential to reshape our understanding of the human mind.

Chapter 6

The Quantum Brain Hypothesis

6.1 Exploring the Orch-OR Theory

The Orch-OR (Orchestrated Objective Reduction) theory is a provocative and controversial hypothesis that posits a quantum basis for consciousness. Proposed by physicist Sir Roger Penrose and anesthesiologist Stuart Hameroff, the theory suggests that consciousness emerges from quantum processes occurring within the microtubules of neurons. In this section, we delve into the Orch-OR theory, exploring its key concepts and implications for our understanding of the enigmatic nature of consciousness.

At the heart of the Orch-OR theory lies the role of *microtubules*, which are tiny protein structures found in the cytoskeleton of neurons. According to Penrose and Hameroff, these microtubules exhibit quantum coherence, a delicate state where quantum superpositions can be maintained over a significant number of particles. They propose that these quantum superpositions play a crucial role in the brain's information processing and the emergence of consciousness.

A fundamental concept in the Orch-OR theory is *objective reduction*, also known as *quantum collapse*. This idea is based on Penrose's earlier work on the nature of quantum measurement. According to the theory, when a certain threshold of quantum superposition is reached within the microtubules, the process of objective reduction occurs. This collapse of the quantum state is suggested to be non-computable and non-algorithmic, thereby introducing a fundamentally non-classical element into brain function.

The Orch-OR theory proposes that these orchestrated objective reductions are not isolated events but occur in synchronized patterns throughout the brain. These orchestrated reductions, happening at a large scale, could be the basis for coherent conscious experiences. The theory attempts to bridge the gap between the seemingly discrete world of quantum mechanics and the macroscopic world of conscious experience.

One intriguing implication of the Orch-OR theory is its potential to explain the phenomenon of *quantum consciousness*. This concept suggests that consciousness is not an emergent property of classical brain processes but rather a fundamental aspect of the universe, existing at the quantum level. If true, this would elevate consciousness to a deeper and more fundamental role in the cosmos.

However, the Orch-OR theory has faced considerable criticism and skepticism from the scientific community. One major concern is the feasibility of quantum coherence within warm and wet biological environments. Maintaining quantum coherence over the necessary timescales in the brain's neural networks presents significant challenges.

Moreover, the nature of objective reduction and its connection to conscious experience is still speculative. The Orch-OR theory does not provide a detailed mechanism for how quantum collapses would give rise to conscious awareness. As such, the theory lacks testable predictions and empirical evidence to support its claims.

Despite the challenges and controversies, the Orch-OR theory has spurred a renewed interest in exploring the potential connections between quantum mechanics and consciousness. It has inspired further research into the role of microtubules and quantum processes in neural information processing.

In conclusion, the Orch-OR theory is a captivating and polarizing hypothesis that proposes a quantum foundation for consciousness. The idea of quantum coherence in microtubules and orchestrated objective reductions offers an intriguing perspective on the nature of consciousness. However, the theory faces significant challenges in terms of its plausibility and empirical support. Nevertheless, it has initiated valuable discussions and investigations into the relationship between quantum mechanics and the enigma of consciousness. As our understanding of both quantum mechanics and neuroscience advances, exploring the Orch-OR theory may shed light on one of the most profound mysteries of the human mind.

6.2 Penrose-Hameroff Model and Quantum Consciousness

The Penrose-Hameroff model is a prominent hypothesis that proposes a quantum mechanical basis for consciousness. It combines the insights of physicist Sir Roger Penrose and anesthesiologist Stuart Hameroff to suggest that certain brain processes involving microtubules give rise to conscious experience. In this section, we delve into the Penrose-Hameroff model, exploring its key components and implications for our understanding of quantum consciousness.

At the core of the Penrose-Hameroff model lies the role of *microtubules*, which are small protein structures found in neurons. These microtubules are proposed to exhibit quantum coherence, a delicate state where quantum superpositions can be maintained over a significant number of particles. Penrose and Hameroff suggest that these quantum superpositions play a critical role in information processing within the brain.

A fundamental concept in the Penrose-Hameroff model is *objective reduction*, also known as *quantum collapse*. This idea is based on Penrose's earlier work on the nature of quantum measurement. According to the model, when a certain threshold of quantum superposition is reached within the microtubules, the process of objective reduction occurs. This non-computable and non-algorithmic collapse of the quantum state is hypothesized to be the basis for conscious experience.

The Penrose-Hameroff model proposes that these objective reductions are fundamental to brain function and consciousness. The orchestrated objective reductions are suggested to be non-random and occur in coherent patterns throughout the brain. These patterns, occurring at a large scale, could account for the unified and coherent nature of conscious experiences.

An intriguing implication of the Penrose-Hameroff model is its potential explanation for the *binding problem*. The binding problem refers to the challenge of how the brain integrates and binds together various sensory and cognitive information into a unified conscious experience. Quantum coherence in microtubules may provide a mechanism for this binding, allowing disparate pieces of information to become entangled and processed as a unified whole.

However, the Penrose-Hameroff model has been met with both interest and skepticism within the scientific community. One major criticism is the feasibility of quantum coherence within warm and wet biological environments. Maintaining quantum coherence over the necessary timescales in the brain's neural networks presents significant challenges.

Moreover, the link between objective reduction and conscious experience is still speculative. While the model suggests a role for quantum collapses in consciousness, it lacks a detailed mechanism for how these collapses give rise to subjective awareness. As such, the Penrose-Hameroff model currently lacks empirical evidence to support its claims.

Despite the criticisms, the Penrose-Hameroff model has sparked renewed interest in the study of quantum consciousness. It has stimulated further research into the role of microtubules and quantum processes in neural information processing and the generation of conscious experience.

In conclusion, the Penrose-Hameroff model is a prominent hypothesis that proposes a quantum basis for consciousness through the role of microtubules and objective reduction. While it faces challenges and skepticism, the model offers an intriguing perspective on the potential connections between quantum mechanics and consciousness. As research in this field continues to evolve, the Penrose-Hameroff model may offer valuable insights into one of the most profound mysteries of the human mind.

6.3 Critiques & Challenges to Quantum Brain Hypothesis

The Quantum Brain Hypothesis, which proposes a quantum mechanical basis for consciousness, has sparked intense debates and discussions within the scientific community. While the hypothesis offers a fascinating perspective on the nature of consciousness, it is not without its share of critiques and challenges. In this section, we delve into some of the key criticisms raised by scientists and philosophers regarding the feasibility and plausibility of the Quantum Brain Hypothesis.

One of the primary critiques centers around the *decoherence problem*. Quantum systems are highly susceptible to environmental interactions, leading to the loss of quantum coherence and the emergence of classical behavior. Maintaining quantum coherence over the necessary timescales in the warm and wet biological environment of the brain is a significant challenge. Critics argue that quantum processes in the brain would quickly succumb to decoherence, rendering any quantum effects irrelevant for cognitive processes.

Another criticism pertains to the *scalability issue*. Quantum computations often require a large number of qubits to achieve their advantages over classical computations. The human brain contains billions of neurons with complex interconnected networks. Emulating quantum computation at such a large scale is currently beyond our technological capabilities. As a result, critics question whether quantum effects could realistically play a significant role in brain function.

The concept of *quantum consciousness* itself has been met with skepticism. While proponents argue that consciousness may have a quantum basis, critics question the need for quantum mechanics to explain conscious experiences. They point to the success of classical theories in describing cognitive processes and argue that quantum effects may be unnecessary to account for consciousness.

Moreover, the lack of direct experimental evidence supporting the Quantum Brain Hypothesis has been a point of contention. While some studies have suggested quantum-like phenomena in biological systems, such as photosynthesis, these findings do not directly demonstrate quantum effects in the brain. The absence of empirical evidence supporting the presence of quantum processes in neural circuits raises doubts about the validity of the hypothesis.

Another challenge involves the *identity of the quantum substrate*, if it exists. Proponents of the Quantum Brain Hypothesis have proposed various candidates for the quantum substrate, including microtubules and ion channels. However, there is no consensus on which specific structures, if any, are responsible for quantum effects in the brain.

The lack of testable predictions is another criticism. Critics argue that the Quantum Brain Hypothesis lacks specific and falsifiable predictions that can be experimentally verified or refuted. Without clear predictions, the hypothesis remains speculative and difficult to subject to rigorous scientific scrutiny.

Ethical concerns have also been raised in discussions surrounding quantum consciousness. If consciousness does have a quantum basis, the potential implications for our understanding of personal identity and free will are profound. Critics caution against drawing unwarranted conclusions or using quantum consciousness as a basis for mystical or pseudoscientific claims.

Furthermore, some critics question the overall coherence of the Quantum Brain Hypothesis. They argue that the hypothesis combines quantum mechanics, neuroscience, and consciousness studies without fully integrating them into a unified framework. The lack of a cohesive theoretical framework raises questions about the overall validity and explanatory power of the hypothesis.

In conclusion, the Quantum Brain Hypothesis has faced significant critiques and challenges from the scientific community. The decoherence problem, scalability issues, lack of direct experimental evidence, and the need for testable predictions are some of the major concerns raised. While the hypothesis has sparked valuable discussions and investigations into the potential links between quantum mechanics and consciousness, it remains a highly debated and controversial topic. As research in this field continues to evolve, addressing these critiques and challenges will be essential for a comprehensive understanding of the role, if any, that quantum processes may play in the complex workings of the human brain and consciousness.

Chapter 7

Experimental Evidence and Observations

7.1 Quantum Effects in Biological Systems

The investigation of quantum effects in biological systems has been a topic of increasing interest and research in recent years. While the focus of this book is on the potential quantum nature of brain waves and cognition, it is essential to explore broader evidence of quantum phenomena in biological processes. In this section, we delve into experimental evidence and observations that suggest the presence of quantum effects in various biological processes.

One remarkable example of quantum effects in biology is *quantum coherence* in photosynthesis. In 2007, researchers discovered that certain photosynthetic complexes, such as the Fenna-Matthews-Olson (FMO) complex in green sulfur bacteria, exhibited long-lasting quantum coherence. Quantum coherence in these systems allows for efficient energy transfer and is believed to enhance the efficiency of photosynthesis in capturing sunlight and converting it into chemical energy.

Another intriguing phenomenon is *quantum tunneling* in enzymes. Quantum tunneling is a quantum mechanical process in which particles can pass through energy barriers that, according to classical physics, would be insurmountable. Some enzymes involved in biological reactions, such as hydrogen transfer reactions in enzymes like lipoxygenase, have been found to exploit quantum tunneling to catalyze chemical reactions more efficiently.

The navigation abilities of certain migratory birds have also sparked interest in quantum biology. These birds can sense the Earth's magnetic field and use it for navigation during their long migrations. While the exact mechanism is not fully understood, it has been suggested that quantum effects in certain light-sensitive molecules in their eyes may play a role in their ability to perceive the magnetic field.

Furthermore, recent studies have highlighted the possibility of quantum processes in *DNA*. The idea of quantum biology in DNA involves the consideration of quantum effects in the process of DNA replication and repair. While the full extent and significance of quantum effects in DNA remain a topic of ongoing investigation, the presence of quantum phenomena in such a fundamental biomolecule has implications for our understanding of biological processes at the molecular level.

Some researchers have also proposed the role of quantum mechanics in *olfaction*. The sense of smell relies on the interaction between odorant molecules and olfactory receptors in the nasal cavity. The ability to distinguish a vast array of smells with high sensitivity has led some scientists to explore the possibility of quantum-based mechanisms in olfaction.

While the evidence for quantum effects in biological systems is intriguing, it is essential to approach these findings with caution. The delicate nature of quantum phenomena makes them susceptible to environmental interactions and decoherence. Some researchers have raised concerns about whether these quantum effects are biologically relevant or merely a consequence of experimental artifacts.

Moreover, the existence and importance of quantum effects in biology remain topics of debate within the scientific community. Some researchers believe that classical physics can adequately explain biological processes without invoking quantum mechanics. Others argue that quantum effects may be at play in specific biological phenomena, but their significance in the overall functioning of living organisms is yet to be fully understood.

In conclusion, the exploration of quantum effects in biological systems is a fascinating area of research that has the potential to revolutionize our understanding of life processes. Evidence of quantum coherence in photosynthesis, quantum tunneling in enzymes, and the potential role of quantum mechanics in navigation, DNA, and olfaction, among other areas, has sparked considerable interest. However, the field of quantum biology is still in its early stages, and many questions remain unanswered. As we continue to advance our experimental techniques and theoretical understanding, we may uncover further evidence of quantum phenomena in living organisms and gain deeper insights into the quantum nature of life.

7.2 Experiments on Quantum Coherence in Brain Tissue

The investigation of quantum coherence in brain tissue has been a subject of significant interest and scientific exploration in recent years. In this section, we delve into some of the key experiments that have been conducted to explore the presence of quantum coherence in the brain and its potential implications for brain function and cognition.

One of the pioneering experiments in this field was conducted by researchers at the University of California, Santa Barbara, who investigated quantum coherence in microtubules within neurons. Microtubules are cylindrical structures found throughout the cytoplasm of cells, including neurons, and are composed of tubulin protein subunits. In this study, the researchers used a technique called two-photon fluorescence microscopy to probe the quantum properties of tubulin. The results indicated that tubulin exhibited characteristics of quantum coherence, raising the possibility of quantum effects in brain tissue.

Another significant experiment involved the study of *quantum entanglement* in brain tissue. Entanglement is a fundamental quantum phenomenon in which two or more particles become connected in such a way that the state of one particle is correlated with the state of another, regardless of the distance between them. Researchers at the University of Alberta used quantum optics methods to examine the possibility of entanglement in brain tissue. While the results were intriguing, the issue of environmental decoherence in the warm and wet biological environment remains a challenge for preserving entangled states over significant distances.

Additionally, quantum effects have been explored in the context of *magneto-reception* in certain animals. Magneto-reception refers to the ability of some organisms to detect the Earth's magnetic field and use it for navigation and orientation. A study conducted at the University of Tokyo investigated whether a protein called cryptochrome, which is believed to be involved in magneto-reception, could be influenced by quantum coherence. The researchers found that quantum coherence in cryptochrome might be responsible for the sensitivity of certain birds to the Earth's magnetic field.

Furthermore, experiments involving *ion channels* have also provided insights into the potential for quantum effects in brain tissue. Ion channels are protein structures that facilitate the movement of ions across cell membranes, playing a crucial role in neuronal signaling. Research at the University of Surrey revealed that ion channels may exploit quantum tunneling to enhance their efficiency in ion transport, suggesting the possibility of quantum processes in neuronal communication.

While these experiments provide intriguing evidence of quantum effects in brain tissue, it is essential to interpret the results with caution. The brain is a complex and highly dynamic organ, and the delicate nature of quantum coherence makes it susceptible to environmental interactions and decoherence. The warm and wet biological environment poses challenges for preserving quantum effects over relevant timescales.

Moreover, the interpretation of experimental results in the context of consciousness and cognition remains a topic of ongoing investigation. While quantum effects at the cellular level are fascinating, the question of whether they play a significant role in brain function and consciousness is still a subject of debate.

In conclusion, experiments on quantum coherence in brain tissue have opened up exciting avenues of exploration in the study of the brain's potential quantum nature. The investigation of microtubules, entanglement, magneto-reception, and ion channels has provided valuable insights into the presence of quantum phenomena in brain tissue. However, many questions remain unanswered, and the field of quantum neuroscience is still in its early stages. As technology and research techniques continue to advance, we may gain deeper insights into the role, if any, that quantum coherence plays in brain function and cognitive processes. Understanding the interplay of quantum mechanics and brain waves could lead to a revolutionary shift in our comprehension of the mysteries of consciousness and the human mind.

7.3 Quantum Phenomena & Cognition: Debate Continues

The relationship between quantum phenomena and cognition remains a subject of intense debate and speculation in the scientific community. While some researchers propose that quantum mechanics plays a crucial role in brain function and cognitive processes, others remain skeptical, favoring classical explanations for brain activity and consciousness. In this section, we explore the arguments on both sides of the debate and examine the challenges of investigating the potential quantum nature of cognition.

Proponents of the *quantum cognition* hypothesis argue that quantum mechanics offers a more comprehensive framework for understanding complex cognitive processes. They point to the phenomenon of *quantum superposition,* where a system can exist in multiple states simultaneously, as a potential explanation for the parallel processing and combinatorial nature of human thought. According to this view, the brain's ability to hold multiple possibilities in superposition may account for creativity, problem-solving, and decision-making.

An example often cited in the context of quantum cognition is the *quantum decision theory* model. It suggests that decision-making involves a quantum-like process, wherein options are represented as probability amplitudes that interfere and collapse during the decision-making process. Some studies have shown that this quantum decision model can provide a better fit to human decision-making data compared to classical models.

On the other hand, skeptics of the quantum cognition hypothesis argue that classical explanations can account for cognitive processes without invoking quantum mechanics. They contend that macroscopic biological systems, such as the brain, are too warm and wet for delicate quantum coherence to persist and influence cognitive functions. Moreover, they question the relevance of quantum effects at the macroscopic level, where classical physics has proven to be highly successful in describing natural phenomena.

The challenge of *quantum decoherence* is a central point of contention in the debate. Decoherence refers to the loss of quantum coherence due to interactions with the environment. Critics argue that the warm and noisy biological environment rapidly destroys delicate quantum states, making it unlikely for quantum effects to have any significant impact on cognition.

Another important aspect of the debate is the issue of *quantum measurement* and its role in consciousness. The process of measurement in quantum mechanics involves the collapse of the wave function, leading to definite outcomes. Some researchers speculate that consciousness may arise from the process of quantum measurement, where the act of observation by a conscious observer leads to the collapse of quantum states. However, this idea remains highly controversial and lacks empirical support.

Moreover, the challenge of *quantum complexity* adds to the complexity of the debate. The brain is an intricate and highly interconnected network of billions of neurons, making it difficult to isolate and study quantum effects at the neuronal level. Experimental investigations in neuroscience often involve macroscopic measurements that may overlook subtle quantum phenomena if they exist at all.

Despite the ongoing debate, the investigation of quantum phenomena in cognition continues to be a fascinating and active area of research. New experimental techniques and theoretical frameworks are constantly emerging to explore the potential quantum nature of brain function. The interplay between quantum mechanics and cognitive processes raises profound questions about the fundamental nature of consciousness and the mind-body relationship.

In conclusion, the debate over quantum phenomena and cognition reflects the complexity of understanding the human mind. Proponents argue that quantum mechanics offers a more comprehensive explanation for cognitive processes, while skeptics favor classical explanations that have been successful in describing macroscopic phenomena. The challenges of quantum decoherence, measurement, and complexity present significant obstacles to investigating the quantum nature of cognition. As the field of quantum neuroscience continues to evolve, we may gain deeper insights into the mysteries of consciousness and the mind, ultimately advancing our understanding of the remarkable interplay between quantum mechanics and brain waves.

Chapter 8

Quantum Brain Waves and Altered States of Consciousness

8.1 Meditation and Brain Waves: A Quantum Perspective

Meditation has been practiced for thousands of years as a means to achieve mental clarity, emotional balance, and spiritual insight. In recent years, scientific research has shed light on the neural mechanisms underlying meditation and its effects on brain waves. This section explores the fascinating connection between meditation, brain waves, and the potential role of quantum processes in altered states of consciousness.

At the core of many meditation practices is the cultivation of *mindfulness*, a state of heightened awareness and presence in the present moment. Mindfulness meditation has been shown to elicit changes in brain wave patterns, particularly in the *alpha* and *theta* frequency bands. Alpha waves, with frequencies between

8 to 12 Hz, are associated with relaxed and calm mental states. Theta waves, ranging from 4 to 8 Hz, are linked to deep relaxation and meditative states. The increase in alpha and theta waves during meditation is believed to signify a shift in cognitive processing and an entry into a more contemplative state of mind.

Quantum theory offers a unique perspective on the altered states of consciousness experienced during meditation. Some researchers propose that the observed changes in brain wave patterns could be attributed to *quantum coherence* in neural networks. According to this view, meditation may facilitate the generation of coherent quantum states in the brain, leading to the observed shifts in brain wave frequencies.

Moreover, the concept of *quantum entanglement* has been invoked to explain the sense of interconnectedness and oneness often reported by experienced meditators. Entanglement suggests that particles can become instantaneously correlated, regardless of distance, once they have interacted in the past. Similarly, some researchers speculate that meditation may induce a form of entanglement at the level of consciousness, allowing meditators to experience a profound sense of unity with the world around them.

The study of *quantum information processing* in the brain is another avenue of exploration in the context of meditation. Quantum information processing involves the manipulation and transmission of quantum information in biological systems. Some theorists propose that the brain may employ quantum computation to process and integrate information more efficiently during meditation, contributing to the altered states of consciousness observed in experienced meditators.

While the connection between quantum mechanics and meditation is intriguing, it is essential to approach these ideas with caution. The brain is an incredibly complex system, and the relationship between brain waves, consciousness, and quantum processes is not yet fully understood. Furthermore, the warm and noisy biological environment poses challenges for preserving delicate quantum coherence, making it difficult to establish a direct link between quantum effects and meditation experiences.

Additionally, individual differences in meditation practices and experiences further complicate the interpretation of research findings. Different meditation techniques may elicit distinct brain wave patterns, and meditators may report varying subjective experiences, making it challenging to draw universal conclusions.

Despite these challenges, the investigation of meditation from a quantum perspective opens up exciting possibilities for understanding consciousness and the mind. Quantum mechanics offers a framework that may help elucidate the mechanisms underlying altered states of consciousness, but more research is needed to bridge the gap between quantum theory and the complexities of brain function.

In conclusion, the relationship between meditation, brain waves, and quantum processes is a fascinating area of research that holds the potential for profound insights into consciousness and altered states of mind. The observed changes in brain wave patterns during meditation and the reported experiences of interconnectedness and unity highlight the intriguing interplay between quantum phenomena and cognitive processes. As scientific understanding and research techniques continue to advance, we may gain deeper insights into the profound effects of meditation on brain function and the remarkable interplay between quantum mechanics and brain waves in the exploration of altered states of consciousness.

8.2 Psychedelics and the Brain: Quantum Enhancements

The study of psychedelics, such as psilocybin, LSD, and DMT, has seen a resurgence in recent years, with researchers delving into their effects on brain function and consciousness. These substances have been used for millennia in traditional rituals and spiritual practices, often claimed to induce altered states of consciousness and mystical experiences. In this section, we explore the fascinating connection between psychedelics, brain activity, and the potential for quantum enhancements in cognition.

Psychedelics are known to interact with the brain's serotonin receptors, particularly the 5-HT2A receptors, leading to altered neural activity and changes in consciousness. Some studies using functional magnetic resonance imaging (fMRI) have shown that psychedelics disrupt the default mode network (DMN), a brain network involved in self-referential thoughts and mind-wandering. Disruption of the DMN is thought to be related to the dissolution of ego boundaries and the sense of interconnectedness often reported during psychedelic experiences.

The effects of psychedelics on brain waves have also been investigated. Studies have revealed increases in *alpha* and *theta* waves during psychedelic experiences, similar to those observed in meditation. The shift in brain wave frequencies suggests a change in cognitive processing and the potential for altered states of consciousness induced by these substances.

The intriguing question arises: could psychedelics enhance cognitive functions through quantum mechanisms? Some researchers propose that the psychedelic experience involves a temporary enhancement of *quantum coherence* in the brain. Quantum coherence refers to the state where quantum systems, such as groups of neurons, become synchronized and entangled, enabling the brain to process information more efficiently.

Moreover, psychedelics may also facilitate *quantum tunneling* in neural networks. Quantum tunneling allows particles to traverse energy barriers that would be insurmountable in classical physics. In the context of the brain, quantum tunneling could enable the transfer of information between brain regions that are not typically strongly connected. This enhanced communication might lead to novel insights and creative thoughts.

The *entropic brain hypothesis* is another intriguing theory in the study of psychedelics. This hypothesis posits that psychedelics temporarily increase the brain's entropy or randomness, allowing for a more diverse range of potential neural configurations. In turn, this increased diversity may contribute to the richness and unpredictability of psychedelic experiences.

However, it is essential to approach the idea of quantum enhancements in psychedelic experiences with caution. The brain is an extraordinarily complex system, and the exact mechanisms underlying psychedelic effects are not fully understood. While some studies support the idea of increased brain wave coherence during psychedelic experiences, the link to quantum phenomena remains speculative.

Moreover, the safety and potential therapeutic benefits of psychedelics require careful consideration. While some research suggests that psychedelics may have therapeutic potential for mental health conditions, such as depression and post-traumatic stress disorder, their use can also lead to adverse psychological effects in vulnerable individuals.

The legality and social stigma surrounding psychedelics have limited research opportunities, hindering a comprehensive understanding of their effects on the brain. As regulations loosen and scientific interest grows, we may gain deeper insights into the fascinating interplay between psychedelics, brain activity, and the potential for quantum enhancements in cognition.

In conclusion, the study of psychedelics and their effects on the brain offers a captivating glimpse into altered states of consciousness and the potential for quantum enhancements in cognition. The observed changes in brain wave patterns and disruptions in brain networks during psychedelic experiences underscore the complexity of brain function. While the idea of quantum coherence and tunneling in the brain is tantalizing, more research is needed to fully explore the quantum aspects of psychedelic experiences. As the field continues to advance, we may unravel the mysteries of psychedelic-induced altered states of consciousness and gain a deeper appreciation for the remarkable interplay between quantum mechanics and brain waves.

8.3   Near-Death Experiences and Quantum Tunneling

Near-death experiences (NDEs) have long fascinated scientists and intrigued the general public. These profound encounters often involve a sense of leaving one's physical body, moving through a tunnel, and entering into a realm of light and transcendence. While NDEs have been attributed to various physiological and psychological factors, some researchers speculate that quantum tunneling may offer an intriguing explanation for the mystical aspects of these experiences.

Quantum tunneling is a quantum mechanical phenomenon where particles can pass through energy barriers that would be classically impossible to surmount. In the context of near-death experiences, some theorists propose that the brain may undergo a form of quantum tunneling during times of extreme stress or trauma. The brain, on the verge of shutting down due to reduced blood flow and oxygen, may enter a quantum state where information processing becomes highly efficient.

One proposed mechanism involves *quantum coherence* in the brain's microtubules. Microtubules are protein structures found in neuronal cells that have been suggested as potential sites for quantum processes in the brain. During an NDE, the brain's neurons may enter a state of enhanced coherence, allowing for the transfer of quantum information across larger spatial scales.

In the context of the tunnel often described in NDEs, quantum tunneling might provide an alternative perspective on the sensation of traveling through a passageway or corridor. As the brain's neural networks undergo quantum processes, the sense of movement through a tunnel-like structure could arise, contributing to the vivid and otherworldly nature of the experience.

While the idea of quantum tunneling in NDEs is intriguing, it remains a speculative hypothesis. The brain's quantum properties, if they exist, are incredibly delicate and difficult to preserve in the warm and noisy environment of living organisms. Additionally, the subjective and highly variable nature of near-death experiences makes them challenging to study scientifically.

Some researchers argue that NDEs could be attributed to the brain's attempt to make sense of the sensory and emotional overload during moments of crisis. The tunnel imagery and feelings of peace and transcendence might arise as the brain processes and integrates fragmented information in an altered state of consciousness.

Moreover, cultural and religious beliefs can significantly shape the interpretation of near-death experiences. In cultures where tunnel imagery is prevalent in narratives of afterlife journeys, individuals who experience NDEs may be more likely to report tunnel-like sensations. These sociocultural factors add another layer of complexity to the interpretation of these extraordinary events.

Despite the challenges in studying NDEs, they offer a unique window into the nature of consciousness and the human experience. Whether arising from quantum processes or not, the profound impact of NDEs on individuals' lives cannot be ignored. Many people who have undergone NDEs report transformative changes, such as a newfound appreciation for life, reduced fear of death, and increased spiritual or existential insights.

In conclusion, the exploration of near-death experiences and their potential connection to quantum tunneling opens up intriguing avenues for understanding altered states of consciousness. While the idea of quantum coherence and tunneling in the brain is captivating, it requires further investigation and empirical evidence. Whether arising from quantum phenomena or not, NDEs remain a profound and deeply personal phenomenon, challenging our understanding of consciousness and the mysteries of human existence. As scientific methods advance, we may gain deeper insights into the nature of these extraordinary experiences and their implications for our understanding of the brain, consciousness, and the cosmos.

Chapter 9

Implications and Future Directions

9.1 Quantum Brain Interfaces and Neurotechnology

Recent advances in quantum mechanics and neuroscience have opened up exciting possibilities for the development of quantum brain interfaces and neurotechnologies. These cutting-edge technologies aim to establish direct communication between the human brain and quantum systems, allowing for unprecedented levels of cognitive enhancement and brain-computer interfacing. In this section, we will explore some of the key concepts and potential implications of quantum brain interfaces in the field of cognitive exploration.

One of the fundamental challenges in brain-computer interfacing is the need for high-bandwidth communication between the brain and external devices. Conventional neurotechnologies, such as electroencephalography (EEG) and functional magnetic resonance imaging (fMRI), have limitations in terms of data transfer rates. Quantum brain interfaces, leveraging the principles of quantum entanglement and superposition, hold the promise of ultrafast and high-fidelity communication between the quantum computer and the brain.

The concept of *quantum brain mapping* is an exciting prospect in this area. By encoding brain states into quantum bits (qubits) and exploiting the principles of quantum entanglement, it may be possible to represent complex brain processes in quantum computational space. This quantum representation could enable a more comprehensive understanding of brain function and cognition, leading to breakthroughs in neuroscience and cognitive science.

Moreover, quantum brain interfaces could revolutionize brain-machine interfaces (BMIs) for individuals with neurological disorders or disabilities. By establishing direct quantum-level communication with the brain, it might be possible to restore lost sensory functions or even augment cognitive abilities. For example, quantum brain interfaces could enable blind individuals to "see" through direct brain stimulation, converting visual information into quantum-encoded signals that the brain can interpret.

However, the development of quantum brain interfaces also raises ethical and philosophical questions. As neurotechnologies advance, issues of privacy, consent, and potential misuse become critical concerns. The ability to access and manipulate brain states at the quantum level brings with it profound ethical implications, necessitating a careful balance between scientific progress and responsible use of these technologies.

Furthermore, the security of quantum brain interfaces is paramount. As direct communication occurs between the brain and external quantum systems, safeguards against malicious attacks or unauthorized access must be in place. Quantum encryption and other security measures will be essential to protect the integrity and privacy of brain data.

The integration of quantum brain interfaces with artificial intelligence (AI) and machine learning opens up new avenues for cognitive enhancement and personalized medicine. Quantum AI algorithms may be developed to analyze quantum brain data and provide insights into individual brain function and cognitive processes. This personalized approach could lead to tailored treatments for neurological conditions and mental health disorders.

In addition to medical applications, quantum brain interfaces could also find use in novel forms of human-computer interaction and virtual reality experiences. Imagine a virtual reality system that directly interfaces with your brain's quantum states, creating a seamless and immersive experience beyond the capabilities of current technology.

However, the realization of quantum brain interfaces is not without challenges. Quantum computing is still in its early stages, and building robust and scalable quantum systems for brain interfacing remains a formidable task. Additionally, our understanding of the brain's quantum properties is limited, and more research is needed to fully comprehend the intricacies of quantum cognition.

In conclusion, quantum brain interfaces and neurotechnologies hold tremendous potential for advancing our understanding of the brain and enhancing human cognition. By leveraging the principles of quantum mechanics, these technologies offer exciting possibilities for brain-computer interfacing, cognitive enhancement, and medical applications. However, ethical considerations, security, and the need for further research and development must be carefully addressed as we venture into this fascinating frontier of neuroscience and quantum mechanics.

9.2 Ethical Considerations of Quantum Mind Research

The exploration of the intersection between quantum mechanics and the human mind raises important ethical considerations. As we delve into the realm of quantum mind research, it becomes crucial to address the potential implications and ethical challenges that arise from investigating the mysteries of consciousness through the lens of quantum mechanics.

One of the primary ethical concerns in quantum mind research is the privacy and confidentiality of individuals' thoughts and mental states. While brain imaging technologies like EEG and fMRI are already capable of providing valuable insights into brain activity, quantum brain interfaces have the potential to access thoughts and cognitive processes at a much more intimate level. Ensuring the privacy and security of individuals' mental states is of paramount importance to protect against unauthorized access or misuse of this highly sensitive information.

Moreover, as quantum brain interfaces advance, the question of informed consent becomes increasingly complex. Participants in quantum mind research may not fully comprehend the intricacies of quantum mechanics and the potential implications of participating in such studies. Ethical guidelines must be in place to ensure that participants are adequately informed about the nature of the research and the potential risks and benefits involved.

Another ethical consideration is the potential for cognitive enhancement through quantum technologies. While the idea of augmenting human cognition through quantum brain interfaces is enticing, it also raises questions about fairness and equality. Will these technologies create disparities between individuals who can afford cognitive enhancements and those who cannot? Ensuring equitable access to such technologies will be essential to avoid exacerbating existing social inequalities.

Additionally, the integration of quantum mind research with artificial intelligence and machine learning raises concerns about the implications of enhanced cognitive abilities. Will enhanced cognitive capabilities lead to a "post-human" society, where individuals with cognitive enhancements have advantages over those without them? Ethical frameworks must be established to address the societal and philosophical implications of these potential scenarios.

As quantum brain interfaces evolve, there is a need to establish international standards and regulations to govern their use. The potential applications of these technologies in military, intelligence, and commercial domains necessitate clear ethical guidelines to prevent misuse and abuse. Collaborative efforts between scientists, ethicists, policymakers, and other stakeholders are vital to create a comprehensive ethical framework for quantum mind research.

The ethical implications of quantum mind research also extend to questions about the nature of consciousness itself. Does quantum mechanics truly play a significant role in shaping consciousness, or are there more fundamental aspects of consciousness that remain to be understood? Addressing these fundamental questions requires a careful balance between scientific exploration and philosophical reflection.

The potential for unintended consequences is another ethical consideration. As we venture into uncharted territories of quantum mind research, there may be unforeseen ethical dilemmas that emerge. Flexibility and adaptability in ethical guidelines will be crucial to respond to novel challenges that arise in this rapidly evolving field.

It is also essential to consider the broader societal impact of quantum mind research. Public understanding and perception of these technologies play a vital role in shaping ethical discourse. Ethical education and public engagement are necessary to foster informed discussions and democratic decision-making on the implications of quantum mind research.

In conclusion, quantum mind research holds great promise for unraveling the mysteries of consciousness and cognition. However, ethical considerations are paramount to ensure the responsible and beneficial development of these technologies. Safeguarding privacy, informed consent, equitable access, and international regulations are among the key ethical principles that should guide quantum mind research. Ethical reflection and dialogue are essential as we navigate the ethical complexities of this fascinating field, ensuring that quantum mind research contributes positively to our understanding of the human mind and the enhancement of human well-being.

9.3 The Road Ahead: Quantum Cognition and Beyond

As we near the conclusion of this cognitive exploration into the fascinating realm of quantum mechanics and brain waves, the road ahead is filled with both promise and uncertainty. The journey into the intersection of quantum cognition and neuroscience has illuminated new pathways for understanding the mysteries of the human mind and consciousness. In this final section, we reflect on the key findings and implications of our exploration and speculate on the future directions that lie ahead.

Our investigation into quantum brain waves has revealed the intricate interplay between quantum phenomena and cognitive processes. From the superposition of brain states to the potential role of quantum entanglement in neural communication, the quantum perspective has offered unique insights into the dynamic nature of cognition. However, many questions remain unanswered, and the pursuit of understanding the quantum nature of the mind is far from over.

One of the crucial takeaways from this journey is the need for interdisciplinary collaboration. The field of quantum cognition requires the synergy of expertise from quantum physics, neuroscience, psychology, computer science, and philosophy. Collaborative research efforts can bridge gaps in knowledge, fuel innovation, and lead to groundbreaking discoveries.

As we delve deeper into quantum cognition, the development of novel experimental paradigms becomes essential. Innovative experimental designs that integrate quantum measurements with brain imaging techniques will be pivotal in uncovering further evidence for quantum effects in the brain. By harnessing advanced technologies and computational tools, researchers can explore complex brain processes in unprecedented detail.

Moreover, the integration of quantum mechanics with other cutting-edge technologies, such as artificial intelligence and quantum computing, presents an exciting frontier. Quantum-inspired AI algorithms may revolutionize cognitive modeling, leading to more accurate and sophisticated models of human cognition. The computational power of quantum computers holds the potential to simulate intricate brain dynamics and offer new avenues for understanding brain function.

Beyond the realm of basic research, the application of quantum cognition to practical domains is another promising direction. Quantum-inspired brain-inspired computing has the potential to enhance pattern recognition, data processing, and decision-making in AI systems. These quantum-enhanced technologies may pave the way for transformative advancements in fields ranging from healthcare to finance.

However, as we forge ahead, it is crucial to remain cautious about overhyping the potential of quantum cognition. The human brain is an exceptionally complex organ, and quantum effects, if they exist, may only play a small role in overall cognitive processes. Keeping expectations grounded in empirical evidence is vital to maintaining scientific rigor.

Furthermore, the ethical considerations that we explored in the previous section are of paramount importance as quantum cognition research progresses. Responsible development, ethical guidelines, and transparency are essential to ensure that these emerging technologies are harnessed for the betterment of humanity.

Finally, the road ahead in quantum cognition is not without challenges and potential pitfalls. The scientific community must remain open to healthy skepticism and rigorous scrutiny. As new theories and experimental findings emerge, they must withstand critical examination to withstand the test of time.

In conclusion, the journey into quantum cognition has been an awe-inspiring voyage into the deepest recesses of the human mind and the quantum universe. The exploration of brain waves, quantum phenomena, and their interplay has opened new vistas for scientific inquiry and cognitive exploration. The road ahead holds promise for groundbreaking discoveries, technological advancements, and a deeper understanding of what it means to be human. As we continue this voyage, let curiosity, collaboration, and ethical responsibility be our guiding stars, propelling us into a future where the mysteries of the mind are illuminated by the dazzling light of quantum cognition.

Chapter 10

Conclusion

10.1 Recapitulation of Key Findings

As we approach the conclusion of this enthralling expedition into the world of quantum mechanics of brain waves, it is time to reflect on the key findings that have emerged throughout this cognitive exploration. Our journey has delved deep into the intricate interplay between quantum phenomena and neuroscience, unveiling profound insights into the enigmatic nature of human cognition and consciousness. In this final section, we revisit the essential discoveries and revelations that have shaped our understanding.

At the heart of our exploration lies the profound realization that brain waves, the rhythmic patterns of neural activity, play a fundamental role in the landscape of cognition. From the serene delta waves of deep sleep to the intense gamma waves of heightened awareness, these oscillatory rhythms orchestrate the symphony of our thoughts and experiences, creating the unique tapestry of our consciousness.

One of the most intriguing and thought-provoking discoveries of our journey has been the potential involvement of quantum mechanics in the cognitive symphony. The idea that quantum phenomena, such as superposition and entanglement, may operate at the cellular level of neurons opens up exciting avenues for unraveling the quantum nature of consciousness.

The concept of superposition, where particles can exist in multiple states simultaneously, finds a captivating analogy in cognitive superposition. Quantum cognitive models suggest that our thoughts and perceptions can exist in a state of simultaneous possibilities, accounting for the richness and flexibility of human cognition, where we can entertain multiple perspectives simultaneously.

Entanglement, another profound quantum phenomenon, hints at the interconnectedness of all things in the universe. Our exploration has revealed the intriguing possibility that entanglement may play a role in the coordination of neural networks, facilitating coherent brain activity and efficient information processing.

The measurement problem, a long-standing enigma in quantum mechanics, also finds an intriguing resonance in the context of brain waves. We have delved into the notion that consciousness may have a role in the collapse of quantum states, influencing the outcome of measurements and shaping our subjective experiences.

Our journey has also taken us to the captivating realm of altered states of consciousness induced by meditation and psychedelic substances. The interplay between quantum coherence and altered states offers a new lens through which we can explore the profound transformations of consciousness that occur in these states, transcending ordinary boundaries of perception.

Furthermore, the exploration of near-death experiences has uncovered the potential of quantum tunneling as a mechanism for consciousness to transcend the confines of the physical brain, hinting at the tantalizing possibility of life beyond the boundaries of our corporeal existence.

As we recapitulate these key findings, it is essential to acknowledge the current limitations and challenges in quantum cognition research. The field is in its nascent stages, and much remains to be explored and validated through rigorous experimentation and theoretical advancements.

The complexity of the human brain and the intricate interplay of cognitive processes demand a cautious approach to interpreting the quantum nature of consciousness. While the quantum perspective offers exciting insights, it is crucial to avoid falling into the trap of quantum mysticism, where quantum mechanics is misapplied or misinterpreted in the context of cognitive phenomena.

Looking ahead, the path forward in quantum cognition is filled with tantalizing possibilities. Advanced brain imaging technologies, quantum-inspired AI algorithms, and cutting-edge computational models offer a myriad of avenues for further exploration into the quantum aspects of cognition.

In conclusion, our odyssey into the quantum mechanics of brain waves has been a breathtaking journey into the frontiers of science and human understanding. We have glimpsed the profound connection between the quantum world and the cognitive landscape of the mind. As we bid farewell to this expedition, we do so with gratitude for the insights gained and the mysteries that continue to beckon us onward. May our quest for knowledge and truth endure, fueled by curiosity, guided by reason, and illuminated by the timeless brilliance of human consciousness.

10.2 The Potential Unification of QM and Brain Waves

As we culminate our enthralling voyage through the realm of quantum mechanics of brain waves, a tantalizing prospect emerges on the horizon - the potential unification of quantum mechanics and the enigmatic world of brain waves. Our journey has taken us on a captivating exploration of the interplay between quantum phenomena and neuroscience, revealing deep connections between the fundamental aspects of reality and human cognition. In this concluding section, we contemplate the exciting possibilities that lie ahead, and how the convergence of these two domains could shape the future of science and our understanding of consciousness.

Throughout our expedition, we have encountered remarkable parallels between quantum principles and the dynamics of brain waves. From the quantum nature of cognition to the potential role of entanglement in neural networks, each revelation has brought us closer to the elusive concept of a unified framework that encompasses both quantum mechanics and brain wave phenomena.

The notion of quantum cognition, where cognitive processes are described by quantum probability distributions, hints at a deeper connection between the probabilistic nature of quantum mechanics and the uncertainty inherent in human decision-making and perception. Such a union could offer profound insights into the nature of human thought processes and pave the way for innovative cognitive models.

The exploration of quantum superposition in the context of cognitive states has also provided compelling evidence for the potential overlap between quantum mechanics and brain waves. The coexistence of multiple possibilities in cognitive superposition aligns with the simultaneous existence of quantum states, hinting at a deeper unity between the two domains.

Moreover, our journey has raised intriguing questions about the role of consciousness in the collapse of quantum states. Could our subjective experiences and conscious observations play a pivotal role in the emergence of definite brain wave patterns and the evolution of our mental states? The unification of quantum mechanics and brain waves may offer a promising framework for exploring this profound connection.

One of the most transformative implications of this potential unification lies in the realm of quantum brain interfaces and neurotechnology. Imagine harnessing the power of quantum coherence to develop brain-computer interfaces that can enhance cognitive functions, or creating quantum-inspired neural networks that mimic the intricate processes of the human brain. The convergence of quantum mechanics and neuroscience could lead to groundbreaking advancements in brain-machine interactions.

As we contemplate the uncharted frontiers that await us, it is essential to acknowledge the challenges and uncertainties that lie ahead. The integration of quantum mechanics and brain waves is an ambitious endeavor that demands interdisciplinary collaboration and rigorous scientific inquiry.

Ethical considerations also come into play when exploring the potential applications of quantum brain research. As we delve deeper into the mysteries of the mind, it becomes imperative to address the ethical implications of manipulating cognitive states and consciousness through quantum means.

However, despite the challenges, the potential benefits are equally captivating. The unification of quantum mechanics and brain waves could revolutionize our understanding of human cognition, lead to groundbreaking treatments for neurological disorders, and unlock the secrets of consciousness itself.

In conclusion, our journey through the intricacies of quantum mechanics of brain waves has been a profound odyssey into the mysteries of the human mind and the fabric of reality. As we look to the future, the potential unification of quantum mechanics and brain waves holds the promise of a new era of scientific exploration and a deeper understanding of the very essence of human consciousness.

10.3 The Potential Unification of QM and Brain Waves

As we culminate our enthralling voyage through the realm of quantum mechanics of brain waves, a tantalizing prospect emerges on the horizon - the potential unification of quantum mechanics and the enigmatic world of brain waves. Our journey has taken us on a captivating exploration of the interplay between quantum phenomena and neuroscience, revealing deep connections between the fundamental aspects of reality and human cognition. In this concluding section, we contemplate the exciting possibilities that lie ahead, and how the convergence of these two domains could shape the future of science and our understanding of consciousness.

Throughout our expedition, we have encountered remarkable parallels between quantum principles and the dynamics of brain waves. From the quantum nature of cognition to the potential role of entanglement in neural networks, each revelation has brought us closer to the elusive concept of a unified framework that encompasses both quantum mechanics and brain wave phenomena.

The notion of quantum cognition, where cognitive processes are described by quantum probability distributions, hints at a deeper connection between the probabilistic nature of quantum mechanics and the uncertainty inherent in human decision-making and perception. Such a union could offer profound insights into the nature of human thought processes and pave the way for innovative cognitive models.

The exploration of quantum superposition in the context of cognitive states has also provided compelling evidence for the potential overlap between quantum mechanics and brain waves. The coexistence of multiple possibilities in cognitive superposition aligns with the simultaneous existence of quantum states, hinting at a deeper unity between the two domains.

Moreover, our journey has raised intriguing questions about the role of consciousness in the collapse of quantum states. Could our subjective experiences and conscious observations play a pivotal role in the emergence of definite brain wave patterns and the evolution of our mental states? The unification of quantum mechanics and brain waves may offer a promising framework for exploring this profound connection.

One of the most transformative implications of this potential unification lies in the realm of quantum brain interfaces and neurotechnology. Imagine harnessing the power of quantum coherence to develop brain-computer interfaces that can enhance cognitive functions, or creating quantum-inspired neural networks that mimic the intricate processes of the human brain. The convergence of quantum mechanics and neuroscience could lead to groundbreaking advancements in brain-machine interactions.

As we contemplate the uncharted frontiers that await us, it is essential to acknowledge the challenges and uncertainties that lie ahead. The integration of quantum mechanics and brain waves is an ambitious endeavor that demands interdisciplinary collaboration and rigorous scientific inquiry.

Ethical considerations also come into play when exploring the potential applications of quantum brain research. As we delve deeper into the mysteries of the mind, it becomes imperative to address the ethical implications of manipulating cognitive states and consciousness through quantum means.

However, despite the challenges, the potential benefits are equally captivating. The unification of quantum mechanics and brain waves could revolutionize our understanding of human cognition, lead to groundbreaking treatments for neurological disorders, and unlock the secrets of consciousness itself.

In conclusion, our journey through the intricacies of quantum mechanics of brain waves has been a profound odyssey into the mysteries of the human mind and the fabric of reality. As we look to the future, the potential unification of quantum mechanics and brain waves holds the promise of a new era of scientific exploration and a deeper understanding of the very essence of human consciousness.

Appendix A

Glossary of Quantum and Neuroscience Terms

As we conclude our journey through the captivating realms of quantum mechanics of brain waves, we recognize the significance of a comprehensive glossary that elucidates the terminology encountered in this extraordinary odyssey. This chapter serves as a reference guide, providing concise and accessible explanations of the key terms and concepts from both quantum mechanics and neuroscience. Our endeavor is to demystify the technical language and facilitate a deeper understanding of the interplay between these two intriguing fields.

Quantum Mechanics Terms

Quantum Superposition: A fundamental principle of quantum mechanics where a quantum system can exist in multiple states simultaneously until measured, represented mathematically by a linear combination of state vectors.

Entanglement: A quantum phenomenon where two or more particles become correlated in such a way that the state of one particle is dependent on the state of another, regardless of distance.

Wavefunction: A mathematical description in quantum mechanics that defines the behavior of a particle as a probability amplitude, giving the probability of finding the particle in a specific state.

Observables: Physical properties of a quantum system that can be measured, such as position, momentum, and energy.

Quantum Measurement: The process of obtaining information about a quantum system by making an observation, which collapses the system's wavefunction to a definite state.

Quantum Coherence: The state of a quantum system when its wavefunctions are in phase, allowing for interference and maintaining a stable superposition.

Quantum Tunneling: A quantum phenomenon where a particle can pass through a potential barrier that classical mechanics would consider impenetrable.

Quantum Decoherence: The process by which a quantum system interacting with its environment loses its coherence and enters a classical-like state.

Schrödinger's Cat: A thought experiment illustrating the concept of superposition, where a cat inside a sealed box is both alive and dead until the box is opened and the cat is observed.

Quantum Entanglement in Brain Waves: The hypothesis that entanglement may play a role in the communication and synchronization of neural activity in the brain.

Neuroscience Terms

Neuron: The fundamental unit of the nervous system, a specialized cell responsible for transmitting information through electrical and chemical signals.

Synapse: The junction between two neurons where chemical neurotransmitters are released to transmit signals from one neuron to another.

Neurotransmitters: Chemical messengers that transmit signals across synapses, influencing neural communication and brain function.

Neural Oscillations: Brain waves generated by synchronous neural activity, associated with different cognitive states and functions.

Electroencephalogram (EEG): A non-invasive technique to record electrical activity in the brain, capturing neural oscillations and brain wave patterns.

Magnetoencephalography (MEG): A non-invasive brain imaging technique that measures the magnetic fields produced by neural activity.

Functional Magnetic Resonance Imaging (fMRI): A brain imaging technique that detects changes in blood flow to identify active brain areas during specific tasks or mental processes.

Consciousness: The state of awareness and subjective experience of the world, the most profound enigma in neuroscience and philosophy.

Neural Correlates of Consciousness (NCC): The neural processes and brain activity that correlate with conscious experiences.

Neural Plasticity: The brain's ability to reorganize and adapt its structure and function in response to experience and learning.

Default Mode Network (DMN): A network of brain regions associated with self-referential thoughts, mind-wandering, and introspection.

Brain-Computer Interface (BCI): Technology that enables direct communication between the brain and external devices, often used for medical and research purposes.

Cognitive Neuroscience: The interdisciplinary study of the neural mechanisms underlying cognitive processes and behavior.

Neuroethics: The ethical implications of neuroscience research and its applications, encompassing topics like brain privacy, cognitive enhancement, and consciousness.

As we conclude this glossary, we recognize that the rich tapestry of quantum and neuroscience terminology represents only a fraction of the vast intellectual landscape that awaits further exploration. The journey into the intricacies of quantum mechanics of brain waves is ongoing, and as science advances, so does our understanding of the profound mysteries that unite the quantum and cognitive realms.

References

Here is a compilation of resources and research papers that serve as valuable references for the book "Quantum Mechanics of Brain Waves: A Cognitive Exploration":

A.1 Books

1. *Principles of Quantum Mechanics* by R. Shankar

2. *Quantum Mechanics and Path Integrals* by Richard P. Feynman and Albert R. Hibbs

A.2 Scientific Journals

- *Nature Neuroscience*

- *Physical Review Letters*

- *Journal of Neuroscience*

- *Trends in Cognitive Sciences*

- *Frontiers in Human Neuroscience*

These resources cover various aspects of quantum mechanics, neuroscience, and the intersection of quantum phenomena and cognitive processes. They can serve as valuable references for the book "Quantum Mechanics of Brain Waves: A Cognitive Exploration."

Acknowledgements

I would like to express my sincere gratitude to all those who have directly or indirectly helped make this project a reality.

I would like to extend my heartfelt gratitude to the readers, the entire scientific community, and my friends and family for their unwavering support and encouragement throughout the creation of this book, "Quantum Mechanics of Brain Waves: A Cognitive Exploration."

To the readers, your curiosity and interest in the fascinating interplay of quantum mechanics and brain processes have inspired me to delve deeper into this subject and present my findings in this work. It is my sincere hope that this book will serve as a valuable resource and spark meaningful discussions and explorations in the fields of quantum cognition and neuroscience.

To the scientific community, I am grateful for the collective efforts of researchers, scholars, and professionals who have contributed to the vast body of knowledge in quantum mechanics and neuroscience. Your groundbreaking research and insights have paved the way for the exploration of new frontiers in understanding the human mind and consciousness.

To my friends and family, your unwavering support and belief in my abilities have been the driving force behind the completion of this book. Your encouragement during moments of doubt and celebration of every milestone have been a constant source of motivation.

In conclusion, this book would not have been possible without the collective efforts and support of all those mentioned above. I dedicate this work to the inquisitive minds, passionate researchers, and individuals who seek to unravel the mysteries of the human brain and the quantum universe. Thank you all for being a part of this journey.

Thank you all for being a part of this journey and for your commitment to advancing mathematical knowledge and understanding.

With immense gratitude,

N.B Singh